Thin Films

Nicoleta Nedelcu

Thin Films

Processes and Characterization Techniques

 Springer

Nicoleta Nedelcu
Department of Deformable Media and Ultrasonics
Romanian Academy
Institute of Solid Mechanics
Bucharest, Romania

ISBN 978-3-031-06615-3   ISBN 978-3-031-06616-0   (eBook)
https://doi.org/10.1007/978-3-031-06616-0

© The Editor(s) (if applicable) and The Author(s), under exclusive license to Springer Nature Switzerland AG 2023

This work is subject to copyright. All rights are solely and exclusively licensed by the Publisher, whether the whole or part of the material is concerned, specifically the rights of translation, reprinting, reuse of illustrations, recitation, broadcasting, reproduction on microfilms or in any other physical way, and transmission or information storage and retrieval, electronic adaptation, computer software, or by similar or dissimilar methodology now known or hereafter developed.

The use of general descriptive names, registered names, trademarks, service marks, etc. in this publication does not imply, even in the absence of a specific statement, that such names are exempt from the relevant protective laws and regulations and therefore free for general use.

The publisher, the authors, and the editors are safe to assume that the advice and information in this book are believed to be true and accurate at the date of publication. Neither the publisher nor the authors or the editors give a warranty, expressed or implied, with respect to the material contained herein or for any errors or omissions that may have been made. The publisher remains neutral with regard to jurisdictional claims in published maps and institutional affiliations.

This Springer imprint is published by the registered company Springer Nature Switzerland AG
The registered company address is: Gewerbestrasse 11, 6330 Cham, Switzerland

Preface

Thin-film coatings, which have been gradually developing since the 1950s, have found widespread application in many fields of science and technology over the years. An optical coating consists of a succession of alternating thin layers (optical thicknesses are comparable to the wavelength of the incident light). The new optical materials obtained have a completely new property.

The subject is of interest to the scientific world and has motivated researchers worldwide to look for numerous methods of evaluating thin films, including both destructive and non-destructive methods.

This book has highlighted the technological processes for obtaining thin films using different deposition methods. This study aims to identify the structural, morphological, and crystalline properties required with the help of microscopy and X-ray techniques. Some general notions and methods of obtaining thin layers with various advantages and disadvantages are highlighted and are classified according to their properties.

Models based on the Swanepoel method were developed using transmittance, absorbance, and reflectance spectra. The experimental data recorded by a spectrophotometer can be used calculate the absorption coefficient, bandgap, Urbach energy, optical conductivity, and dielectric constant.

UV-VIS-NIR and IR spectrum ellipsometry (SE) are non-destructive indirect optical techniques that allow the characterization of thin films, surfaces, and interfaces used to determine the thickness of thin films, optical constants, bandgap, and electrical parameters (mobility, conductivity, and carrier density).

Ellipsometry doesn't directly measure the film thickness or optical constants. The analysis of ellipsometry is performed using an optical model. This model is an approximate structure of the samples and includes the order of layers for this material, optical constants, and the thickness of the layers. The experimental data can compare the result obtained by spectroscopy in the visible and ultraviolet fields.

Bucharest, Romania

Nicoleta Nedelcu

Acknowledgments

This book arose from studying the chalcogenide layer while working in the project Improve. I want to thank Dr. V. Chiroiu, without whose efforts this book could not have been written, my family, my husband for supporting me, and M. Dulgheru for graphics and artwork.

I would like to thank the staff at Springer, in particular Michael Luby and Amrita Unnikrishnan, for their help and support.

Contents

1 Types of Optical Coating Systems 1
   1.1 Introduction 2
   1.2 Types of Inhomogeneities for Optical Materials 7
   1.3 Uniformity of Optical Coating 9
      1.3.1 Determination of the Uniformity on Aspherical
           Surfaces in the Planetary System Geometry 9
      1.3.2 Plane Support 17
      1.3.3 Spherical Dome 19
      1.3.4 Pyramidal and Conical Dome 20
      1.3.5 Planetary System Geometry 23
      1.3.6 Uniformity Screens 26
   References 31

2 Deposition Methods, Classifications 33
   2.1 Thin Layers Method 33
   2.2 Thin Layer Obtaining by Thermal Evaporation Method 34
   2.3 Thermal Evaporation by Laser Ablation 36
   2.4 Obtaining Thin Layers by Thermal Spraying 37
      2.4.1 Detonation Gun Spraying 39
      2.4.2 Electric Arc Spraying Process 40
      2.4.3 Plasma Spraying 41
   2.5 Electrochemical Methods 43
   2.6 Chemical Vacuum Deposition Method (CVD) 44
   References 45

3 Vacuum Deposition 47
   3.1 Vacuum Thin Film Deposition Installations 47
   3.2 Evaporation Devices 49
   3.3 Important Materials for Evaporation 53

ix

	3.4	Thickness Measurement Methods	54
		3.4.1 Capacitive Method	56
		3.4.2 Inductive Method	56
		3.4.3 Resonant Quartz Method	57
		3.4.4 Gravimetric Method	58
	References		58
4	**Morpho-structural Characterization**		**61**
	4.1	Methods for Characterizing Thin Layers	61
	4.2	Structural and Morphological Analysis	62
	4.3	Neutron Diffraction (ND)	63
	4.4	Microscopy Techniques	66
		4.4.1 Atomic Force Microscopy (AFM)	66
		4.4.2 Scanning Electron Microscopy (SEM)	69
		4.4.3 Transmission Electron Microscopy (TEM)	74
	References		86
5	**Optical Analysis and Chemical Properties**		**89**
	5.1	Study of Optical and Chemical Properties	90
		5.1.1 UV-VIS Spectroscopy	90
	5.2	UV-VIS-NIR and IR Spectrum Ellipsometry (SE)	102
		5.2.1 The Effective Medium Approximation (EMA) or EMT (Effective Medium Theory)	106
		5.2.2 Modeling Data	107
	5.3	Study of Chemical Properties	111
		5.3.1 Infrared Spectroscopy-IR (FT-IR)	111
		5.3.2 Raman Spectroscopy	113
	References		118
Index			**121**

Chapter 1
Types of Optical Coating Systems

Abstract Since the 1950s, gradually developing thin-film coatings have been widely applied in many fields of science and techniques. Spectacular applications of thin layers are made in microelectronics, electronics high frequencies, laser techniques, automata, computers, filters, unique optical mirrors, radiation detectors, sensors, various optoelectronic devices, and many devices used in industry. Uniformity is the most important parameter that needs to be adjusted in production to obtain a quality material. This chapter presents calculations for the geometric coefficient in vacuum and the deposition installation to determine the uniformity thickness in plane and aspherical geometry systems.

Keywords Refractive index · Dispersion equations · Standard dispersion · Sellmeier dispersion · Optical material · Uniformity of optical coating · Geometry system · Plane support · Spherical geometry · Plane geometry · Aspherical surface · Geometric coefficient · Spherical dome · Planetary system · Uniformity screen · Conical dome

Thin-film coatings, which have been in development since the 1950s, have found extensive application across various fields of science and technology. Spectacular applications of thin layers are made in microelectronics, electronics high frequencies, laser techniques, automata, computers, filters, unique optical mirrors, radiation detectors, sensors, various optoelectronic devices, and many devices used in industry. Uniformity is the most important parameter that needs to be adjusted in production to obtain a quality material. This chapter presents calculations for the geometric coefficient in vacuum and the deposition installation to determine the uniformity thickness in plane and aspherical geometry systems.

© The Author(s), under exclusive license to Springer Nature Switzerland AG 2023
N. Nedelcu, *Thin Films*, https://doi.org/10.1007/978-3-031-06616-0_1

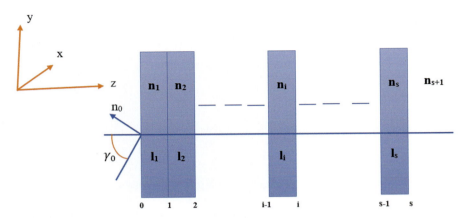

Fig. 1.1 Schematic representation of a multilayer consisting of thin layers

1.1 Introduction

Thin layers are widely applied in many fields of science and techniques. The most spectacular applications of thin layers are made in microelectronics, electronics high frequencies, laser techniques, automata, computers, etc. Filters, unique optical mirrors, radiation detectors, sensors, various optoelectronic devices, and many devices used in industry are built with thin layers [1–3].

The optical coatings are used to modify optical surfaces' optical, electrical, mechanical, and chemical properties. An optical coating consists of a succession of thin layers (optical thicknesses are comparable to the wavelength of incident light) of alternating optical materials. The scheme of an optical coating is presented in Fig. 1.1 and consists of s thin layers, surrounded at the two edges by massive media.

The layers are numbered from left to right, and we assume that the incident radiation is linearly polarized, flat, monochrome, and infinitely extended. Assume that the multilayer is made up of thin layers bordered by infinitely extended plane-parallel interfaces; the layers are isotropic. The complex refractive n_c index completely describes the optical properties of each layer:

$$n_c = n_i + ik_i \tag{1.1}$$

where n_i is the refractive index of the medium, k_i is the extinction coefficient of material $i = \sqrt{(-1)}$, and geometric thickness l_i.

Suppose that the refractive index of the incident medium is nonabsorbent. The mathematical formalism used to determine the spectral characteristics of coatings is a matrix; each layer is characterized by a matrix, which depends on the phase thickness of the layer i, φ_i [4, 5].

1.1 Introduction

$$M_i = \left[\cos \varphi_i \frac{i}{n_i} \sin \varphi_i i n_i \sin \varphi_i \cos \varphi_i \right] \tag{1.2}$$

$$\varphi_i = \frac{2\pi}{\lambda} n_i l_i \tag{1.3}$$

The relation used in the calculation is:

$$\begin{bmatrix} E_{i-l} \\ B_{i-l} \end{bmatrix} = \left[\cos \varphi_i \frac{i}{n_i} \sin \varphi_i i \sin \varphi_i \cos \varphi_i \right] \begin{bmatrix} E_i \\ B_i \end{bmatrix} \tag{1.4}$$

E_i, B_i are the aptitudes of the electric and magnetic field vectors. The factors of reflection R, transmission T, jump in refraction ς and transmission ξ and absorption A_i in a i layer is calculated as follows:

$$R = \left| \frac{E_0 - (B_0/n_0)}{E_0 + B_0/n_0} \right|^2 \tag{1.5}$$

$$T = \frac{4n_{s+1}}{n_0 |E_0 + (B_0/n_0)|^2} \tag{1.6}$$

$$\varsigma = \arctan \left[\frac{E_0 - (B_0/n_0)}{E_0 + (B_0/n_0)} \right] \tag{1.7}$$

$$A_i = \frac{\Im \begin{bmatrix} E_{i-l} \\ B_{i-l} \end{bmatrix} - \Im \begin{bmatrix} E_i \\ B_i \end{bmatrix}}{\frac{n_0}{4} |E_0 + (B_0/n_0)|^2} \tag{1.8}$$

The following relationships of the refractive index $\widetilde{n}_i$ used for oblique incidence are the following:

$$\text{Polarization } p \to \widetilde{n}_i = \frac{n_i}{\cos \phi_i} \tag{1.9}$$

$$\text{Polarization } s \to \widetilde{n}_i = n_i \cos \phi_i \tag{1.10}$$

$$\text{Polarization } s \text{ and } p \to \widetilde{\phi}_i = \phi_i \cos \phi_i \tag{1.11}$$

where

$$\cos \phi_i = \left(\frac{\sqrt{a_i^2 + b_i^2} + a_i}{2} \right)^{\frac{1}{2}} - \left(\frac{\sqrt{a_i^2 + b_i^2} - a_i}{2} \right)^{\frac{1}{2}}, \tag{1.12}$$

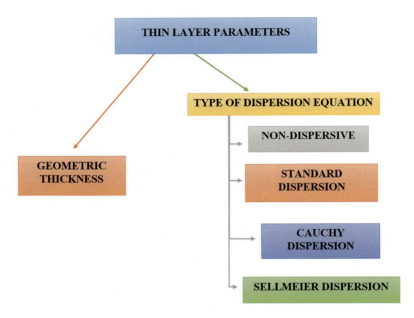

Fig. 1.2 Thin layer parameters characterization diagram

$$a_i = 1 + \left(k_i^2 - n_i^2\right)\left(\frac{n_0 \sin \gamma_0}{n_i^2 + k_i^2}\right)^2, \qquad (1.13)$$

$$b_i = -2n_i k_i \left(\frac{n_0 \sin \gamma_0}{n_i^2 + k_i^2}\right)^2. \qquad (1.14)$$

The thin layers are characterized by the parameters presented in Fig. 1.2.

The light enters the nonabsorbing homogeneous material and reflection and refraction occur at the boundary surface. The refractive index n is the ratio between the velocity of light in vacuum c and the medium ν.

$$n = \frac{c}{\nu} \qquad (1.15)$$

The refractive index is a measure for the strength of deflection occurring at boundary surface at refraction of the light beam, described by the Snell's law equation:

$$n_i \sin \alpha_i = n_j \sin \alpha_j \qquad (1.16)$$

The most common characteristic to characterize an optical glass is to calculate the refractive index values depending on the function of wavelength. Dispersion is explained to the molecular structure of matter by applying the electromagnetic

1.1 Introduction

wave. The bound changes vibrate at the frequency of the incident wave, if the electromagnetic wave impinges on an atom or molecule [6, 7]. Nondispersive medium is a homogeneous and isotropic medium, and the wave velocity is constant, A_1 for all wavelengths, the dispersion equation is described by the relation:

$$n = A_1 \tag{1.17}$$

An accurate description of the optical properties of glass is achievable to the characterized optical glass through the refractive index. The Abbe number alone is insufficient for high-quality optical systems that need the aid of relative partial dispersion.

The difference between the refractive index values at 486.1 and 656.3 nm is called the principal dispersion. The Abber number is defined as:

$$v = \frac{n_{589.3}(\lambda) - 1}{n_{486.1}(\lambda) - n_{656.5}(\lambda)} \tag{1.18}$$

but in general, the formulation is defined as:

$$v = \frac{n_{\text{center}} - 1}{n_{\text{short}} - n_{\text{long}}} \tag{1.19}$$

where n_{center}, n_{short}, n_{long} are the refractive indices of the material at three different wavelengths. The equation defines the relative partial dispersion $D_{a,b}$ for the wavelength a and b:

$$D_{a,b} = \frac{n_a(\lambda) - n_b(\lambda)}{n_{486.1}(\lambda) - n_{656.3}(\lambda)} \tag{1.20}$$

Abbe demonstrated that the linear relationship applies to the majority of glasses, which are called "normal glasses":

$$D_{a,b} \approx X_{a,b} + Y_{a,b} \cdot v \tag{1.21}$$

$Y_{a,b}$, $X_{a,b}$ are specific constants for the given relative partial dispersion.

The Cauchy formula gives another way to express the dependence of the refractive index on the wavelength:

$$n(\lambda) = A + \frac{B}{\lambda^2} + \frac{C}{\lambda^4} \tag{1.22}$$

where A, B, C are dependent on material and constants specifically obtained for different wavelengths. The Cauchy equation is applied at normal incidence and is used in the visible spectrum region for many glasses. The original expression [7] described a series in terms of wavelength λ or frequency ω of light:

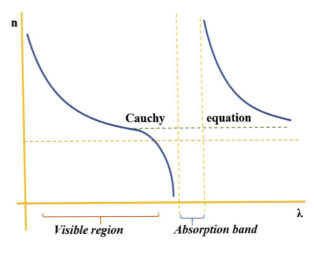

Fig. 1.3 The graphic representation of the Cauchy equation

$$n = n_1 + n_2\omega^2 + n_4\omega^4 \tag{1.23}$$

The graphic representation of the Cauchy equation for refractive index versus wavelength is in Fig. 1.3.

Suppose the refractive index is measured in a transparent substance or bulk into the infrared region of the spectrum. In that case, the dispersion curve begins to show marked deviations from the Cauchy equation. The deviation curve presents a significant discontinuity in the dispersion curve as wavelength increases. Sellmeier[1] [8] obtained an equation representing an excellent improvement for the refractive index calculated by the Cauchy equation. The Sellmeier equation is derived from the classical dispersion theory. It allows the description of the refractive index over the total transmission with one set of data and calculates intermediate values accurately. The empirical relation between the refractive index n and wavelength λ is given by:

$$n^2 = 1 + \sum_{i=1}^{3} \frac{A_i \lambda^2}{\lambda^2 - \lambda_i^2} \tag{1.24}$$

where λ_i, A_i is called the Sellmeier coefficient, which is determined by fitting this expression in experimental data. The Sellmeier formula has more terms in a similar form but generally can be neglected.

Sellmeier's equation represents the dispersion curve better than the Cauchy curve, as shown in Fig. 1.4.

Thin layers can be obtained by mechanical, chemical well by condensation from the gas phase. The gas phase condensation method entered many processes by which thin layers can be obtained: thermal evaporation in a vacuum, cathodic spray, plasma

---

[1] In 1871, Sellmeier obtained the equation.

1.2 Types of Inhomogeneities for Optical Materials

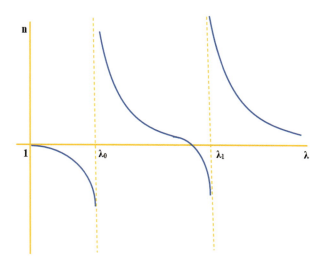

Fig. 1.4 The graphic representation of the Sellmeier equation

spray, vapor reaction (growth epitaxial), oxidation reactions, etc. The most commonly used process for obtaining thin layers through gas phase condensation is thermal evaporation in a vacuum or an inert atmosphere; this process has the advantages that make it accessible and, as such, quite widespread. As the evaporating material, we used zinc selenide crystals. The obtained layers were utilized to investigate the electrical, optical, and photoelectric properties of *ZnSe*. Getting high-quality layers and correlating conditions of submission with their structure were the objectives of our research. Another important task was to study the influence of heat treatment on structures and the surface morphology of the samples obtained.

1.2 Types of Inhomogeneities for Optical Materials

The profile of the refractive index varies by the type of inhomogeneity chosen for the material. The values of refractive index n and extinction coefficient k are obtained for experimental data measure in transmission or reflection, using a spectrophotometer or ellipsometer. The refractive index is evaluated at position f within the layer, relative to the interface from the substrate. This facilitates the simulation of the photometric control process during the manufacture of the layer.

The types of homogeneity description form are:

- Homogeneous material,
- Linear inhomogeneous (Fig. 1.5), described by the relation:

Fig. 1.5 Interface diagram of a linear inhomogeneous material

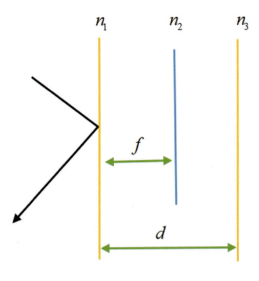

Fig. 1.6 Interface diagram of a slope inhomogeneous material

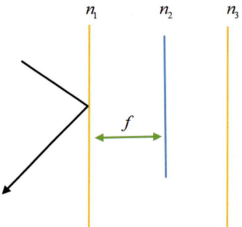

$$n(\lambda) = n_1(\lambda) + [n_2(\lambda) - n_1(\lambda)]\frac{f}{d} \qquad (1.25)$$

- Inhomogeneous slope (Fig. 1.6), described by relation:

$$\begin{aligned} n(\lambda) &= n_1(\lambda) + p(\lambda) \times f \\ p(\lambda) &= p_1 + a_1\lambda + a_2\lambda^2 + \frac{a_3}{\lambda^2} \end{aligned} \qquad (1.26)$$

where p_1, a_1, a_2, a_3 are constants given by the material.

1.3 Uniformity of Optical Coating

The optical coatings depositions on the surfaces of different optical components do not lead to a uniform coverage (geometric thicknesses of the layers that make up the optical coating do not have the same value at each point of the surface). When studying the spectral behavior of optical coatings, we must also know the uniformity of the optical surfaces of the optical system. The uniformity is further analyzed on the thin layers [9]. The surfaces can be described with quadric layers obtained in chemical vacuum deposition where we have sources (crucibles from which the thin layer material is evaporated). It is necessary to know the physical properties. The surface on which the coating is deposited has a rotational motion or a planetary motion.

1.3.1 Determination of the Uniformity on Aspherical Surfaces in the Planetary System Geometry

The geometry of evaporation in a vacuum deposition installation has a planetary motion type. Figure 1.7 illustrates the photometric type thin layer thickness control system. Two reference systems are represented:

- One is a fixed axis (O, c_1, c_2, c_3) related to the installation;
- The second is a mobile axis related to the quadric surface, (O', c'_1, c'_2, c'_3).

The system (O', c'_1, c'_2, c'_3) has originated on the unit vector c_3 at the distance h from O, and the unit vector c'_1 is rotated in a plane parallel with the plane (c_1, O, c_2). The unit vector c'_3 is parallel to the symmetry axis of the quadric surface. The h_1 is the distance between the quadric symmetry axis surface and the unit vector c'_3. The angle between the unit vectors c_3 and c'_3 is $\pi - \alpha$, where α is constant. The reference system (O', c'_1, c'_2, c'_3) is rotated around the c_3 with angular velocity ω.

The quadric rotation is around the unit vector c'_3 with angular velocity $\omega_1 = k \cdot \omega$, where k is constant. The position of the vector $\vec{S}$ is (s_1, s_2, s_3) to the evaporation source. The control of the thickness in thin layers is done by measuring the reflection factor and/or transmission of the optical coating made on a test bulk. The thickness geometry of the thin layers at a point on the optical surface component is called the geometric coefficient of the point F, noted with c_g and is given by the relation:

$$c_g = \frac{g_F}{g_b} \tag{1.27}$$

where

g_F – is the geometric thickness of the layer in F,
g_b – is the geometric thickness of the layer in bulk.

Fig. 1.7 Representation of the geometry evaporation in a planetary system

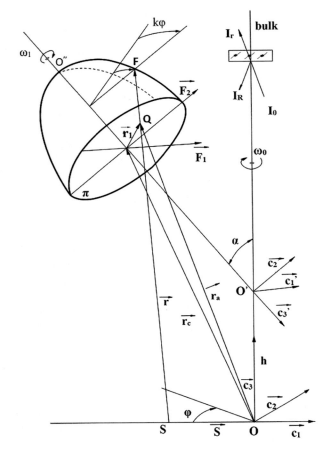

The geometric coefficient will characterize the uniformity of the layers. It is necessary to consider a small sphere of the elementary surface, dS, where the evaporated material is at a constant evaporation rate m (kg/s) in all directions. Such a source of evaporation is called the point source. The amount of material [10] that passes through the solid angle $d\omega$ in any direction in the unit of time is given by:

$$dm = \frac{m}{4\pi} d\omega \qquad (1.28)$$

Assuming that the material is evaporated in the direction of the vapor beam and deposited on the inclined surface element dS_1 at an angle ϕ (as depicted in Fig. 1.8), we can determine the thickness geometry and the quantity of material deposited on the surface element using the following equation:

1.3 Uniformity of Optical Coating

Fig. 1.8 The diagram of rotation

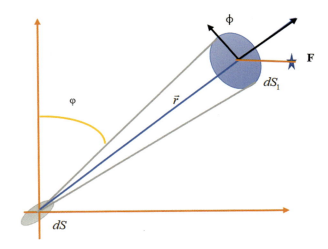

$$d\omega = \frac{\cos \phi}{r^2} dS_1 \qquad (1.29)$$

The rate of material deposition on the element surface dS_1 is:

$$dm = \frac{m \cos \phi}{4\pi r^2} dS_1 \qquad (1.30)$$

and for the case of the point source, the rate of material deposition is:

$$dm = \frac{m \cos \varphi \cos \phi}{\pi r^2} dS_1. \qquad (1.31)$$

We assume that the evaporated material has a density ρ (kg/m³) and the thickness of the layer condensed in the unit of time (deposition rate) ς (m/s), then the volume of material deposited on dS_1, ςdS_1 is $dm = \rho \cdot q \cdot \varsigma dS_1$. The deposit rate for the point source can be written as:

$$q = \frac{m \cos \phi}{4\pi \rho r^2} \qquad (1.32)$$

and for the plane source is:

$$q = \frac{m \cos \phi \cos \varphi}{\pi \rho r^2} \qquad (1.33)$$

For simple geometric evaporations (the surface on the deposited material can be a plane, or dome spherical, fixed or in rotation), the geometric thickness distribution on the surface can be done analytically [9, 10]. Then the surface is more complicated, and a more complex movement takes place; the thickness deposition of the thin layers cannot also be deduced analytically. In the latter case, numerical methods are used [11]. Evaluating the geometric thickness in a point F is done numerically, assessing the rate of deposition of the material in the successive points that describe the movement. The evaluation of the deposition rate must determine the position vector $\vec{r}$ that unites on the surface of the source and the point F the normal $\vec{n}$ at the quadric surface at point F and regular at the source (which is constant). In cases, a plane source, the rate deposition at a point F is proportional to:

$$q_F = a \frac{\cos \phi_F \cos \varphi_F}{r_F^2} \tag{1.34}$$

where a is a constant, ϕ_F is the angle between the normal to the quadric surface in point F, and the vector $\vec{r}_F$, φ_F is the angle between the normal to the surface of the source and the vector $\vec{r}$. The size of the source can be neglected, considering it an elementary source. There is dependence between the movement of rotation and the angular movement. The position of the point F during the movement depends only on the angle of rotation around the unit vector c_3, $\varphi = \omega \cdot t$, where t is the time. It is considered that at $t = 0$ the point F, the unit's vectors c_1, c_3, c_1', c_2' and source S are in the same plane. Relationship (1.33) can be written as:

$$q_\varphi = a \frac{\cos \gamma_\varphi \cos \delta_\varphi}{r_\varphi^2} \tag{1.35}$$

where φ is the angle traveled during t by the unit vector c_1'. The thickness g_F on the quadric area at a point F made for a complete rotation is:

$$g_F = a \int_0^{2\pi} q_\varphi d\varphi \tag{1.36}$$

The uniformity of the layer in the meridian and the parallel surface containing the point F are studied. We consider the eccentricity of the quadric surface c is the distance between the quadric axis and $O'O''$ axis. For $\varphi = 0, 2\pi, 4\pi, \ldots, c = 0$ and the positions of the point F coincide, then the points on the parallel that contain the points F do not have the same thicknesses. The points on the parallels have the same thicknesses, only if the ratio $k = \frac{\omega_1}{\omega}$ is an irrational number. After a relatively large number of complete rotations, all points from the parallel, at the positions $\varphi = 2\pi k$, $k = 1, 2, 3, \ldots$, pass through the initial position of F. If $c \neq 0$ then, uniformity cannot be obtained in parallel. In the case represented in Fig. 1.8, the point F does not see the source, if the surface is concave, it is possible for certain values of φ, the deposition rate is zero for their positions. For point F to have a line of sight to the source, the vector connecting the source and point F must pass through the opening

1.3 Uniformity of Optical Coating

of the quadric. Analytical calculation of uniformity thin layer on various surfaces can be done only for a small number of special cases. The numerical solution of the problem will be approached [11]. Considering that the piece rotates around c_3 in n_i relatively small steps $\Delta\varphi$, each pass has a constant deposition rate at a point F. The thickness g_F where the point F sees the source achieved after n steps ($n \leq n_i$) are:

$$g_F = \sum_{i=0}^{n} q_i = a \sum_{i=0}^{n} \frac{\cos \gamma_{\varphi_i} \cos \delta_{\varphi_i}}{r_{\varphi_i}^2} \tag{1.37}$$

The thickness g_b of the bulk (in the center of the bulk) is:

$$g_b = n_i a \frac{\cos^2 \gamma}{r_t^2} \tag{1.38}$$

The geometric coefficient for the point F is:

$$c_g = \frac{g_F}{g_b} = \frac{\sum_{i=0}^{n} \frac{\cos \gamma_{\varphi_i} \cos \delta_{\varphi_i}}{r_{\varphi_i}^2}}{n_i \frac{\cos^2 \gamma}{r_t^2}} \tag{1.39}$$

where $n_i \geq n$. The field $\left(O', c_1', c_2', c_3'\right)$ will start to define the quadric plane.

$$Q_F = \sum_{i=0}^{3} \sum_{j=0}^{3} A_{ij} x^i x^j + 2 \sum_{i=0}^{3} B_i x^i + C = 0 \tag{1.40}$$

with $A_{ij} = A_{ji}$. The coordinates $\left(x_1^1, x_1^2, x_1^3\right)$ represent the location of point F on the quadric plane. The plane tangent at a point F to the quadric has the equation:

$$\sum_{i=0}^{3} \sum_{j=0}^{3} A_{ij} x^i x_1^j + \sum_{i=0}^{3} B_i \left(x^i + x_1^i\right) + C = 0 \tag{1.41}$$

It will be written like:

$$\sum_{i=0}^{3} A_i x^i + B = 0 \tag{1.42}$$

A_i – describing the orientation of the normal vector to the tangent plane.

$$n = \sum_{i=0}^{3} A_i c_i' \tag{1.43}$$

In step i, the quadratic is rotated around the unit vector c_3' with the angle $\chi = ik\Delta\varphi$. This can be considered a centro-affine transformation [12] of the matrix.

$$[S_i^j] = [\cos\chi - \sin\chi 0 \sin\chi \cos\chi 0 0 0 1] \tag{1.44}$$

The new coordinates of point a F in the system (O', c_1', c_2', c_3') become:

$$\begin{bmatrix} x^1 \\ x^2 \\ x^3 \end{bmatrix} = [S_i^j] \begin{bmatrix} x_1^1 \\ x_1^2 \\ x_1^3 \end{bmatrix} \tag{1.45}$$

and the coordinates of the normal become:

$$\begin{bmatrix} A_1' \\ A_2' \\ A_3' \end{bmatrix} = [S_i^j] \begin{bmatrix} A_1 \\ A_2 \\ A_3 \end{bmatrix} \tag{1.46}$$

The coordinate system (O', c_1', c_2', c_3') is rotated around the unit vector $\vec{c}_3$ with the angle $\varphi = i\Delta\varphi = \omega t$, and the matrix becomes:

$$[B_i^j] = [\cos\varphi \cos\alpha \sin\varphi \sin\alpha \sin\varphi \sin\varphi - \cos\alpha \cos\varphi - \sin\alpha \cos\varphi 0 \sin\alpha - \cos\alpha] \tag{1.47}$$

As $t = 0$ it was considered that c_1', c_1, c_3 are in the same plane. The coordinates of the point $F(y_1, y_2, y_3)$ are in the same plane in system (O', c_1', c_2', c_3') and become $F(x^1, x^2, x^3)$ in the system (O, c_1, c_2, c_3).

$$\begin{bmatrix} x^1 \\ x^2 \\ x^3 \end{bmatrix} = [B_i^j] \begin{bmatrix} y^1 \\ y^2 \\ y^3 \end{bmatrix} + \begin{bmatrix} 0 \\ 0 \\ h \end{bmatrix} \tag{1.48}$$

and the normal $\vec{N}$ becomes:

1.3 Uniformity of Optical Coating

$$\begin{bmatrix} N^1 \\ N^2 \\ N^3 \end{bmatrix} = [B_i^j] \begin{bmatrix} A_1' \\ A_2' \\ A_3' \end{bmatrix} \tag{1.49}$$

The source S has the coordinates (s_1, s_2, s_3) in the reference system (O, c_1, c_2, c_3). The $\overline{SF}$ distance is:

$$\overline{SF} = |\vec{r}| = \sqrt{(x^1 - s_1)^2 + (x^2 - s_2)^2 + (x^3 - s_3)^2} \tag{1.50}$$

The unit vector F to the vector $\vec{r}$ has the relation:

$$\cos\alpha = \frac{x^1 - s_1}{|\vec{r}|}, \ \cos\chi = \frac{x^2 - s_2}{|\vec{r}|}, \ \cos\gamma = \frac{x^3 - s_3}{|\vec{r}|} \tag{1.51}$$

The angle between $\vec{N}$ and $\vec{F}$ is:

$$\cos\phi_F = \frac{\vec{N}\,\vec{F}}{|\vec{N}||\vec{F}|} = \frac{N^1\cos\alpha + N^2\cos\chi + N^3\cos\gamma}{\sqrt{(N^1)^2 + (N^2)^2 + (N^3)^2}} \tag{1.52}$$

The angle between the normal to the plane of the source and the radius $\vec{r}$ is:

$$\cos\phi_F = \frac{\vec{F}\,\vec{c_3}}{|\vec{F}||\vec{c_3}|} = \cos\gamma \tag{1.53}$$

In cases, the optical coating is deposited on the convex surface and we consider it normal outside the quadric, then point P sees the source only if $\cos\phi_F > 0$. Having all the data, now the next step is to determine the evaporation rate at any time of movement. We are going to test if the vector $\vec{r}$ passes through the quadric opening. Denote the plane containing the opening of the quadric by Π. A basis is defined orthonormal in the plane Π of the vectors $\overrightarrow{F_1}$ and $\overrightarrow{F_2}$, with $\overrightarrow{F_1} \parallel c_1', \overrightarrow{F_2} \parallel c_2'$. The point of intersection of vector $\vec{r}$ with plane Π, denoted as Q, can be expressed using the position vector $\vec{r_a}$ as follows:

$$\vec{r_a} = \vec{r_c} + \vec{r_1} = \vec{r}_c + t_1\overrightarrow{F_1} + t_2\vec{F} = \vec{r_c} + t_1\vec{c_1'} + t_2\vec{c_c'} \tag{1.54}$$

where $\vec{r_c}$ – is the position vector in the center of the quadric opening. The unit vectors $\vec{c_1'}$ and $\vec{c_2'}$ can be expressed in the system (O, c_1, c_2, c_3) as:

$$\vec{c_1'} = \sum_{j=1}^{3} B_1^j \vec{c_j},$$

$$\vec{c_2'} = \sum_{j=1}^{3} B_2^j \vec{c_j}, \qquad (1.55)$$

The coordinates for the point Q became:

$$\begin{bmatrix} a^1 \\ a^2 \\ a^3 \end{bmatrix} = \begin{bmatrix} x_c^1 + t_1 B_1^1 + t_2 B_2^1 \\ x_c^2 + t_1 B_1^2 + t_2 B_2^2 \\ x_c^3 + t_1 B_1^3 + t_2 B_2^3 \end{bmatrix} \qquad (1.56)$$

The position vector $\vec{r_a}$ can also be written as:

$$\vec{r_a} = \vec{S} + t_3 \vec{F} \qquad (1.57)$$

where $\vec{F}$ is the unit vector of the vector $\vec{r}$ and the coordinates of the Q point can be written as:

$$\begin{bmatrix} a^1 \\ a^2 \\ a^3 \end{bmatrix} = \begin{bmatrix} s_1 + t_3 \cos \alpha \\ s_2 + t_3 \cos \chi \\ s_3 + t_3 \cos \gamma \end{bmatrix} \qquad (1.58)$$

Having relations (1.54) and (1.56), we can write:

$$\begin{bmatrix} x_c^1 + t_1 B_1^1 + t_2 B_2^1 \\ x_c^2 + t_1 B_1^2 + t_2 B_2^2 \\ x_c^3 + t_1 B_1^3 + t_2 B_2^3 \end{bmatrix} = \begin{bmatrix} s_1 + t_3 \cos \alpha \\ s_2 + t_3 \cos \chi \\ s_3 + t_3 \cos \gamma \end{bmatrix} \qquad (1.59)$$

$$\begin{cases} B_1^1 t_1 + B_2^1 t_2 - \cos \alpha t_3 = S_1 - r_c^1 \\ B_1^2 t_1 + B_2^2 t_2 - \cos \chi t_3 = S_2 - r_c^2 \\ B_1^3 t_1 + B_2^3 t_2 - \cos \gamma t_3 = S_2 - r_c^3 \end{cases} \qquad (1.60)$$

The condition for the radius to pass through the opening of the quadric radius δ is:

$$t_1{}^2 + t_2{}^2 < \delta^2 \qquad (1.61)$$

This relation is applied when the optical coating is deposited on the concave surface; for the convex surface, the condition is:

1.3 Uniformity of Optical Coating

$$\vec{N} \cdot \vec{F} < 0 \qquad (1.62)$$

Where $\vec{N}$ it is oriented towards the outside of the quadric surface.

Planetary system geometry must contain particulars of evaporation geometries type like:

- plane,
- spherical dome,
- conical and pyramidal dome.

We need these types of evaporation geometries to determine the analytical distribution of the thickness of thin layers. This type of case can verify the algorithm for planetary system geometry.

1.3.2 Plane Support

In these cases, it is necessary to consider the coating deposition in plane support, which is perpendicular to the axis of rotation of the plane (Fig. 1.9).

The geometric thickness deposited on plane support at a point F is positioned in the axis of rotation by radius r, and is given by equation [10]:

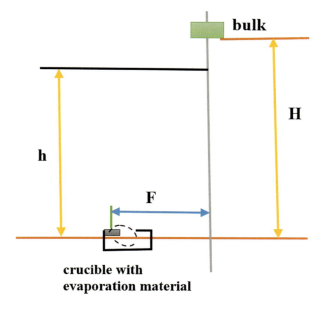

Fig. 1.9 The diagram of evaporation on a support plane

$$g_F = \frac{mh^2\left[2h^2 + (F+r)^2 + (F-r)^2\right]}{2\rho\pi\left\{\left[h^2 + (F+r)^2\right]^{\frac{3}{2}}\left[h^2 + (F-r)^2\right]^{\frac{3}{2}}\right\}} \tag{1.63}$$

The geometric thickness g_b deposited on the bulk plane support is:

$$g_b = \frac{mH^2}{\rho\pi\left(H^2 + F^2\right)^2} \tag{1.64}$$

The geometric coefficient c_F of the point F is given by the relation:

$$c_F = \frac{g_F}{g_b} = \left[\frac{(H^2 + F^2)^2 h^2}{2H^2}\right]\left\{\frac{2h^2 + (F+r)^2 + (F-r)^2}{\left[h^2 + (F+r)^2\right]^{\frac{3}{2}}\left[h^2 + (F-r)^2\right]^{\frac{3}{2}}}\right\} \tag{1.65}$$

1.3.2.1 Calculation Examples for the Geometric Coefficient

We consider the geometry: $H = 590$ mm, $h = 500$ mm, $F = 210$ mm, the geometric coefficient is obtained by applying the relation (1.65). The values are represented in Fig. 1.10.

Figure 1.11 shows the calculation of the geometric coefficient using the algorithm for planetary geometry. The radius is measured in Fig. 1.10 from the center, and in Fig. 1.11, it is measured from the edge to the center. There is an excellent concordance between the two determinations.

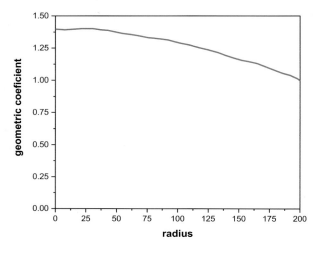

Fig. 1.10 The geometric coefficient calculated for a plane geometry

1.3 Uniformity of Optical Coating

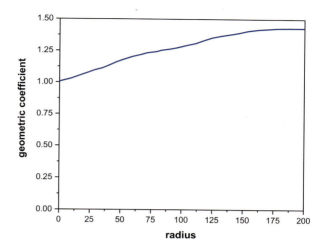

Fig. 1.11 The geometric coefficient is calculated by the algorithm for planetary geometry

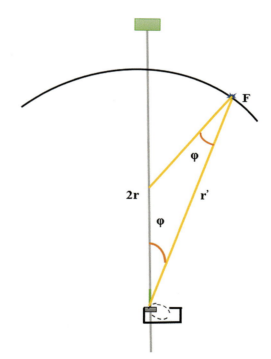

Fig. 1.12 Spherical geometry

1.3.3 Spherical Dome

Spherical dome geometry is described in Fig. 1.12. The radius of the spherical dome is r, the source is at a distance $2r$ to the dome. In the spherical dome, the thickness obtained is constant.

Fig. 1.13 Plane geometry

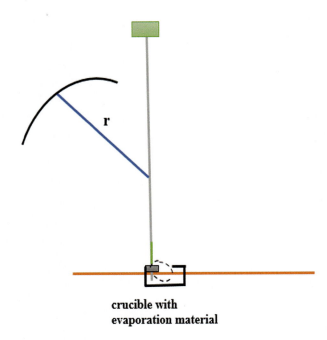

crucible with
evaporation material

The thickness obtained at a point F is:

$$g_F = \frac{m\cos^2\varphi}{\pi\rho r'^2} = \frac{m}{\pi\rho(2r\cos\varphi)^2} = \frac{m}{4\pi\rho r^2} = \text{const} \qquad (1.66)$$

Planetary geometry is derived from spherical geometry and is shown in Fig. 1.13. The surface where the material is deposited has the same radius r. The spherical surface always belongs to spherical dome geometry. In this case, constant thickness is obtained on the spherical surface. The algorithm satisfies this test.

1.3.4 Pyramidal and Conical Dome

Pyramidal domes are frequently used because they allow the introduction of different shapes in the same technological cycle. Netterfield [13] studied the uniformity of thin layers in conical dome geometry and the problem of making screens, fixed and mobile, to increase uniformity in these geometries. For example, the geometry studied by [13], in Fig. 1.14, checked the developed algorithm to see if it gives the same results.

The uniformity is presented in Fig. 1.15 for several inclinations of the conical surface. The results obtained coincide with the results presented in [13].

Next, consider the geometry of the BAK 550 installation, shown in Fig. 1.16.

1.3 Uniformity of Optical Coating

Fig. 1.14 Conical dome-type geometry

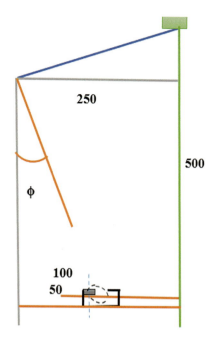

Fig. 1.15 Uniformity for conical dome geometry

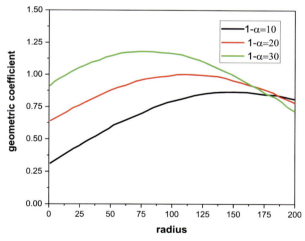

The uniformity obtained in this geometry is represented in Fig. 1.17. The uniformity was determined by deposited thin layers of Ti_3O_5. Depositing SiO_2 layers, a difference was found between the theoretical and the experimental data because the polar distribution of the beam intensity evaporated, for the SiO_2 crucible has deviations from the plane source. This is considered in the calculation algorithm. In the case of evaporation with the electron beam, the polar distribution of the evaporated beam intensity can be described by $\cos^n \phi$, where n varies between 1.0 and 3.0, and depending on the strength of the electron beam [9, 13]. For $n = 0$ we have the case of the point source, and for $n = 1$ we have the case of the plane source.

Fig. 1.16 Spherical dome geometry for the BAK 550 installation

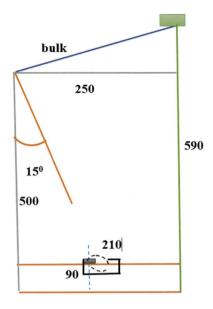

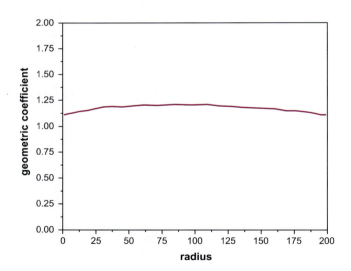

Fig. 1.17 Uniformity geometry for BAK 550

1.3 Uniformity of Optical Coating

1.3.5 Planetary System Geometry

If the axis of revolution coincides with the axis of symmetry of the quadric surface, then the nonuniformity is symmetrical about the axis of quadric symmetry. All points belonging to the intersection of a plane perpendicular to the symmetry axis of the quadric (points on the parallel) must have the same geometric coefficient. Representing the geometric coefficients for the points in the meridian plane, the graphic must be symmetrical to the tip of the quadric. Figure 1.18 presents a planetary geometry that has the equation:

$$\frac{y^2}{2} = -0.32x^2 + 64x \qquad (1.67)$$

All heights are given to the base of the technological chamber. Choosing the field of integration 5 rotations with step $\Delta\Phi = 0.50$ and speed ratio of $k = 2.33$, we obtain the uniformity represented in Fig. 1.19.

It is observed that the graph has small deviations from symmetry. If the graph is not symmetrical, the integration steps (the field integration and speed ratio) were

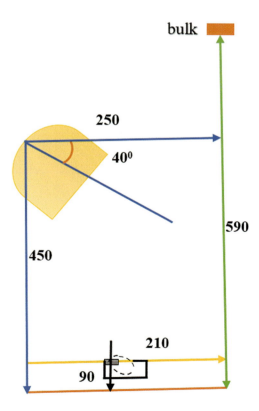

Fig. 1.18 The planetary system geometry

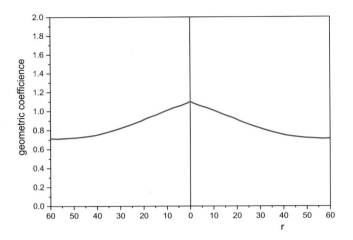

Fig. 1.19 The uniformity obtained by the geometry is presented in Fig. 1.18

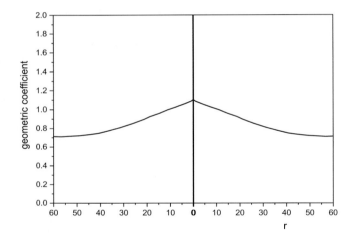

Fig. 1.20 The uniformity obtained for the geometry of Fig. 1.18 with correct integration parameters

chosen incorrectly. Theoretically, the ratio between the speed of revolution and rotation must be an irrational number. In this case, if we increase the integration range to 15 rotations, we obtain the uniformity presented in Fig. 1.20.

If the eccentricity is $c > 0$, the point on the parallel containing the point F does not have the same uniformity. In this case, the uniformity must be studied both along meridians and along with parallels. It considers the geometry of Fig. 1.21, the quadric equation is given by the relation (1.67).

The uniformity along the meridian is studied in Fig. 1.22, and Fig. 1.23 represents the uniformity studied in parallel. When evaluating the uniformity in parallel, it is

1.3 Uniformity of Optical Coating

Fig. 1.21 The planetary geometry with eccentric ($c = 700$ mm)

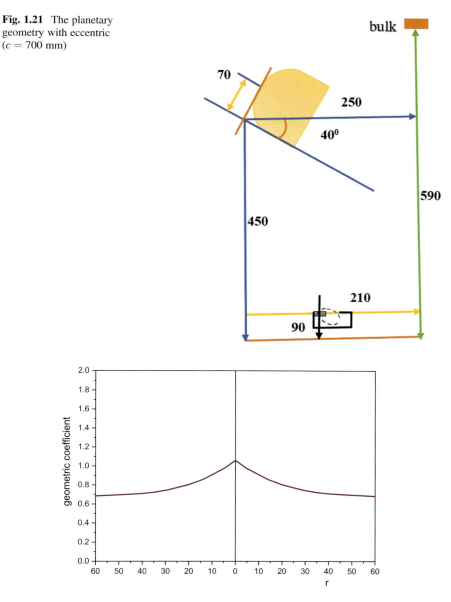

Fig. 1.22 The geometry's uniformity applied to meridian is studied in Fig. 1.21

always necessary to check the domain, and the integration step must be correctly chosen. So, we can see that the meridian's uniformity does not have the same geometric coefficient for a parallel.

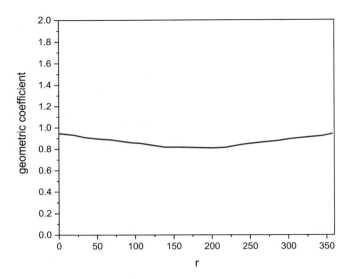

Fig. 1.23 The parallel uniformity for the geometry studied in Fig. 1.21

The uniformity of optical coating deposited on surfaces of the pieces plans, spherical or pyramidal, can be determined using the planetary system's geometry. It is necessary to determine:

- the coordinates of the peak surface;
- the axis angle of symmetry surface (axis of symmetry and axis of rotation are in the same plane).

1.3.6 Uniformity Screens

The constructive parameters of the installation limit the evaporation geometries that can be deposited in a vacuum installation. Even for some geometries of optimized evaporation, the resulting uniformity is not always satisfactory. Take, for example, the case of the evaporation geometry on plane support represented in Fig. 1.24. The geometric thickness of the layer in a point at a distance R from the rotation axis is [10]:

$$\widetilde{g} = \frac{mH^2}{2\rho\pi^2} \int_0^\pi \frac{2\theta}{\left[H^2 + (R+A)^2 - 4RA \sin^2 \frac{\theta}{2}\right]^2} \quad (1.68)$$

For a given geometry there is a point with minimum geometric thickness so that by rotating the point with 180°, the point sees the source throughout the rotation. The

1.3 Uniformity of Optical Coating

Fig. 1.24 The rotation of the plane

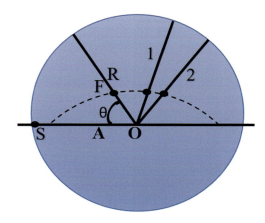

other points have higher thicknesses, as a result of which to reach the same minimum thickness, the points must be rotated by less than 180°.

This is done by adding a screen between the source and the plane for a constant thickness on the plane. The plane of the uniformity screen is parallel to the plane (c_1, O, c_2). In Fig. 1.24 the evaporation geometry perpendicular to the plane is presented. where S is the plane source at a distance A from the rotation axis, $\widetilde{g}_1$ is the minimum thickness, and for each point F, we have a thickness $\Delta\widetilde{g} = \widetilde{g} - \widetilde{g}_1$. If the uniformity screen is at position 1, defined by the angle θ_1, then the screen uniformity size measured on the radius R in point F is given by the position 2 (angle θ_2) with the condition:

$$\Delta\widetilde{g} = \frac{mH^2}{2\rho\pi^2} \int_{\theta_1}^{\theta_2} \frac{2d\theta}{\left[H^2 + (R+A)^2 - 4R\sin^2\frac{\theta}{2}\right]^2} \tag{1.69}$$

In this case, the screen uniformity is not symmetrical. If the screen is symmetrically around position 1, the uniformity screen needs to start at $\theta_1 - \theta_2$ and end at $\theta_1 + \theta_2$:

$$\Delta\widetilde{g} = \frac{mH^2}{2\rho\pi^2} \int_{\theta_1-\theta_2}^{\theta_1+\theta_2} \frac{2d\theta}{\left[H^2 + (R+A)^2 - 4R\sin^2\frac{\theta}{2}\right]^2} \tag{1.70}$$

In the case of symmetrical screens, the position θ_1 is the bisector of the angle defined by the two elementary sources. The relationships (1.69) and (1.70) are used to determine the angle θ_2, if the screen exceeds the range $[0, \pi]$ then in relation (1.69), the integration range will be 0 to 2π.

Fig. 1.25 The pyramidal geometry of the dome

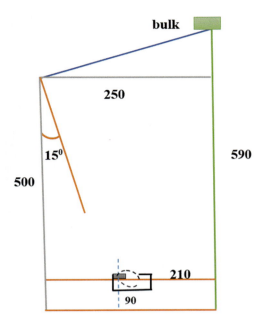

In general, the screen uniformity is put opposite to the elementary source because in that area, the evaporation rate is lower, and the position execution errors of the screen do not lead to significant nonuniformities.

The same reasoning can be done for spherical dome geometries and pyramidal domes [11–13]. Suppose we want to design a screen for growth uniformity of the layers in geometry represented in Fig. 1.25 (the dome pyramidal geometry). The uniformity for this geometry is presented in Fig. 1.26. To ensure uniform deposition of all layer's geometries, the shutter screen of uniformity is considered parallel to the plane (c_1, O, c_2). The screen has a high symmetric 490 mm. It is calculated above and opposite the vapor source (crucible), imposing the geometric coefficient $\tilde{g}_1$. The shapes of the screens are represented in Figs. 1.27 and 1.28.

The shape of the shutter screens for planetary system geometry is a circle. The uniformity screens cannot be determined like other evaporation geometries. Point F does not always have the same path in the right direction screen. That is why the user gives the screen parameters. The screens have the shape shown in Fig. 1.29.

Suppose we have planetary-type geometry, as shown in Fig. 1.25. We find that the maximum thickness is obtained at the top of the quadric and the minimum thickness at the quadric edge. To increase the uniformity, it is necessary to acquire the point where the evaporation rate is high at the top and low at the quadric edge. This area is the opposite of the source.

In Fig. 1.30 representsd the uniformity for three symmetrical screens, positioned at $\theta = 180°$, with apertures 120°, 200°, 280°, and height $H = 350$ mm.

It is observed that for large angular apertures of the screen, the uniformity increases. Despite this, the decrease in the geometric coefficient results in a significant amount of evaporated material being consumed. This makes it challenging to

1.3 Uniformity of Optical Coating

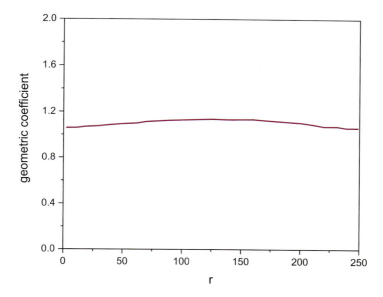

Fig. 1.26 The uniformity is obtained by the pyramidal geometry of the dome

Fig. 1.27 The shape of the uniformity screen placed above the source to obtain $\widetilde{g}_1$

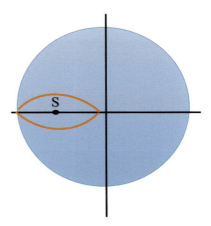

use photometric control for optical coatings in the technological process and requires the use of interference measurement filters that operate far outside the spectral range of the coatings.

Suppose the surface bulk on which it is deposited is convex. In that case, the parameters of planetary geometry must be chosen carefully, considering that material vapors can be incident on the surface at large angles, which makes the layers porous and with low mechanical strength.

Fig. 1.28 The shape of the uniformity screen placed on the opposite side of the source to obtain $\widetilde{\overline{g}}_1$.

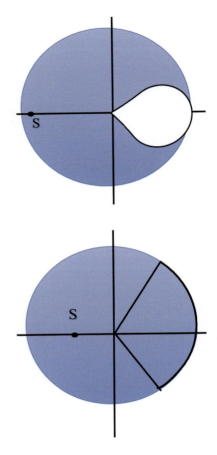

Fig. 1.29 The shape of the uniformity screen is presented for the planetary system

Fig. 1.30 Uniformity for the geometry of Fig. 1.25

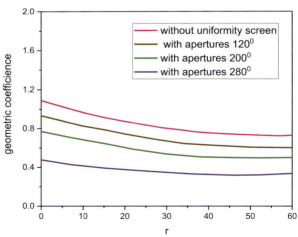

In cases when the surface on which it is deposited is large, it is possible that, during planetary motion, the distance between the vapor source and some F points on the surface be small, which leads to high evaporation rates that can create problems with the properties of thin layers. Throughout the planetary motion, it is ensured that no point on the surface experiences an evaporation rate exceeding a certain threshold, as dictated by the need to achieve thin layers with specific optical and mechanical properties. Uniformity screens can also be used to obtain a nonuniformity imposed.

References

1. L. I. Maissel, R. Glang, *Handbook of Thin Film Technology*, New York, McGraw Hill Book, 1970, pg. 568–625, (1970).
2. В. В. Слуцкая, *Тонкие пленки в технике сверхвысоких частот*, Москва, Советское Радио, 1967, pg.320–419, (1967).
3. I. Spînulescu, *Fizica straturilor subţiri şi aplicaţiile acestora*, Bucureşti, Ed. Ştiinţifică, 1975, pg. 44–62–356–425, (1975).
4. T. Minami, *Present status of transparent conducting oxide thin-film development for Indium-Tin-Oxide (ITO) substitutes*, Thin Solid films 516 (17), 5822–5828, (2008).
5. John L. Vosen, *Thin Film Processes*, Academic Press, pg. 5–95, (1978).
6. H. Bach & N. Neuroth, *The properties of optical glass* Springer pg. 19–164, (1998).
7. R. Jedamzik, S. Reichel, P. Hartmann, *Optical glass with tightest refractive index and dispersion tolerances for high-end optical designs*, SPIE Proceeding 8982–51, (2014).
8. J. Misterik, S. Kasap, H. E. Ruda, C. Koughia, J. Singh, Optical properties of electron materials: fundamentals and characterization, Part A, pg. 47–81, Springer Nature, 2017.
9. U. Petzold, R. Jedamzik, P. Hartmann, and S. Reichel, *V-Block refractometer for monitoring the production of optical glasses*, Proc. SPIE 9628, (2015).
10. L. Holland, *Vacuum Deposition of Thin Films*, Chapman & Hall Ltd. London, pg.92–176 (1960).
11. C. C. Lee, *Making aspherical mirrors by thin-film deposition*, Appl. Opt. 32 (28) pg. 5535–5540, (1995).
12. V. Cruceanu, *Elemente de algebra liniara si geometrie*, Editura Didactica si Pedagogica, Bucuresti, pg. 181–216, (1973).
13. R. P. Netterfield, *Uniform evaporated coatings on rotating conical workholders*, J. Vac. Sci. Technol., **19** (2), Jul./Aug. 1981.

Chapter 2
Deposition Methods, Classifications

Abstract This chapter presents different methods to obtain thin layers specified to the metallurgical and optical industry. In the metallurgical industry, thin layers are used to coat parts that withstand mechanical wear or chemical corrosion. Some general notions and methods are presented with various advantages and disadvantages, highlighted and are classified according to their properties. Several modern installations for obtaining thin layers coatings, which have a high degree of automation and complexity, are revealed.

Keywords Thin layer · Deposition methods · Uniformity layer · Laser beam · Physical and chemical methods · Electrochemical method · Gas phases · Epitaxial growth · High-frequency current · Thermal evaporation method · Laser ablation method · Thermal spraying · Detonation gun spraying · Electric arc spraying · Plasma spraying · Plasma jet · Chemical vacuum deposition (CVD)

This chapter presents different methods to obtain thin layers specified to the metallurgical and optical industry. In the metallurgical industry, thin layers are used to coat parts that withstand mechanical wear or chemical corrosion. Some general notions and methods are presented with various advantages and disadvantages, highlighted and are classified according to their properties. Several modern installations for obtaining thin layers coatings, which have a high degree of automation and complexity, are revealed.

2.1 Thin Layers Method

As we discussed in the previous chapter, the thickness of thin layers must be uniform to control the composition of the layers since the deposition phase. The microstructure of the layers is very important because it is closely related to their optical and magnetic properties. Figure 2.1 shows a schematic of some methods for obtaining thin layer deposits used in the industry.

© The Author(s), under exclusive license to Springer Nature Switzerland AG 2023
N. Nedelcu, *Thin Films*, https://doi.org/10.1007/978-3-031-06616-0_2

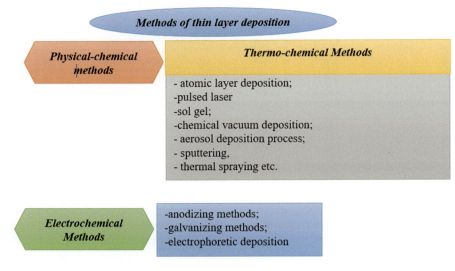

Fig. 2.1 Methods for obtaining thin layers used in industries

2.2 Thin Layer Obtaining by Thermal Evaporation Method

The vacuum thermal evaporation method consists of two main steps:

- Substance evaporation that is deposited,
- Subsequent condensation on the support.

Auxiliary phases accompany these two stages: preparation, clearing substrate, high vacuum, followed by the processing of the condensed layer. The crystalline structure, the physical properties, the composition, and the purity degree obtained by layer depend on gases in the evaporation chamber, etc.

Thin layers can be obtained by mechanical, chemical, or gaseous condensation processes. By mechanical methods [1], relatively thick layers can be obtained, the thinnest foils are obtained by lamination, rarely reaching thicknesses of less than $50 \div 20$ μm. Chemical processes can obtain thin layers with micron thicknesses or much thinner.

In the case of the deposition method by condensation of the gas phases, the process of the thin layer has two phases mentioned above. The epitaxial growth condensation mechanisms differ within certain limits from the growth mechanisms. In obtaining layers by vacuum thermal evaporation or by high-frequency plasma spraying, the method depends on the different growth and structural form mechanisms. The conductors allow for the passage of electric current, either through the substance deposited on them or potentially even from the material evaporated to make them. Methods such as high-frequency current heating, bombardment by ionic or electronic beams, and using electric arcs are also employed. By heating, the substance begins to evaporate before or after the melting temperature.

2.2 Thin Layer Obtaining by Thermal Evaporation Method

Some materials, such as cadmium, zinc, magnesium, some semiconductor compounds, by heating, pass directly from the solid state to the vapor state. Obviously, to obtain a net beam of evaporated substance particles (atoms, molecules), an advanced vacuum must be made in the evaporation chamber. The crystalline structure, the physical properties, the composition, and the degree of purity obtained by the thin layer depend largely on the presence of gases in the evaporation chamber.

In contrast to single-layer deposition, in the binary or more complex depositions, compounds can create various difficulties related to the deviation of the stoichiometric composition of the thin layer or from the initial composition of the evaporated material [2]. However, suitable methods such as the Veksinsky method [3] or the three-temperature method [4] have been possible to obtain thin layers whose composition is not different from the original material.

The evaporation process strongly depends on the temperature of the substrate, the nature and the degree of cleanliness, and other parameters that may vary during the condensation layer. Among them can be mentioned the intrinsic, constant parameters specific to the evaporated substance:

- Melting temperature,
- Sublimation,
- Composition,
- Nature of the material, as well as parameters that characterize the evaporation–condensation process such as:
- Support-evaporator distance,
- Vacuum density atomic (molecular) beam,
- Critical condensation temperature,
- Mobility of atoms on the surface of the support,
- The presence of magnetic or electric fields,
- Radiation,
- Impurities.

The physicochemical properties [5], crystalline structure, adhesion to the substrate, thickness, and stoichiometric composition (in the case of compounds) are all influenced by the parameters mentioned.

The vacuum deposition method has the advantage of achieving a wide range of thin layers at low substrate temperatures, including room temperature.

The disadvantages of the method are the introduction of impurities from the vacuum chamber walls, especially the glass ones (which are absorbed as a result of radiant heating and bombardment with secondary electrons), and the capture of charges in the films deposited due to the substrate [6]. The thermal evaporation method is limited to use volatile materials at moderate temperatures, which do not react with the filament materials (crucible) at the evaporation temperature and do not decompose under the evaporation conditions.

2.3 Thermal Evaporation by Laser Ablation

The technique is called in the literature as "laser ablation" or "pulsed laser deposition" (ablation = the process of removing fine particles from the surface and transporting them through a specific process; pulsed lasers – need to use high power, required pulsating lasers).

The high temperature developed by the laser beam allows the evaporation of hard fusible materials (diamond, ceramic). From the point of view of the operating mechanism, the laser beam irradiates the target made of the material to be deposited, causing, depending on the specific constant of the optical absorption of the surface, its heating to a certain temperature. The temperature is dependent on the power of the laser source, the duration of the pulses, and the diffusivity coefficient of the target surfaces. At a sufficient power level, the surface reaches the temperature necessary for the evaporation of the constituents, and the evaporation occurs instantly. Experimentally, the most useful laser radiation for ablation is that in the ultraviolet field, and the optimal pulse duration is 30 ns.

The high surface temperature results in a broad spectrum of particles: electrons, ions, neutral atoms, and molecules. The continuous action of the laser beam causes the particles above the target to undergo a photoionization process, resulting in the formation of a plasma cloud. Due to the excess of an electron, this plasma Laser thermal ablation evaporation by thermal evaporation laser ablation becomes strongly absorbent for the radiation beam, which leads to the irradiation of its temperatures. An elongation of the plasma cloud takes place, acquiring a specific shape, that of a feather. Elongation decreases the density due to the penetration of radiation to the target surface.

Laser ablation technology is currently used for the deposition of superconducting layers, ceramic materials, and especially for obtaining atomic layers without remote order.

The experimental PLD deposition installation is shown in Fig. 2.2.

The deposition installation contains [7] an excimer laser source that generates a high-brightness laser pulse, entering through a quartz window into the reaction chamber.

Laser pulse energy can be adjusted and monitored by a coherent system consisting of a measuring head and an energy analyzer. The duration of the laser pulse is measured with a detector and viewed by an oscilloscope. The laser beam is focused on the target surface with a cylindrical MgF_2 lens located outside the deposition chamber. The process of heating and cooling the substrate can be controlled with a constant slope by the temperature controller.

The cooling is at the same pressure used during deposition. To eliminate the possibility of any contamination and guarantee the gas's purity during the deposition process, the reaction chamber needs high pressure. To remove residual contamination before pulse application to obtain the deposited layer, several consecutive cleaning pulses must be applied. During the cleaning pulses, a screen is inserted between the target and the collector. The initial ablated substance condenses on the screen, with a high concentration of impurities.

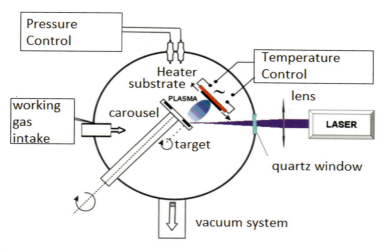

Fig. 2.2 PLD general assembly diagram

In a deposition process, [8] several targets of the same material or different materials can have multi-structures. This option offers the advantage of avoiding the exposure of coatings to the environment during the process. Repeated opening on the deposition chamber may lead to changes in the material already deposited by reactions with oxygen or by adsorption of the molecule on the surface. After deposition, the samples are subjected to a heat treatment in water vapor for 6 hours, heated to the same temperature as that applied during the deposition. After-deposition heat treatment is to improve the crystallinity of the coatings and reconstitute/preserve the stoichiometry of the compound. All heating and cooling processes are done with a constant ramp to avoid thermal stress due to sudden temperature changes, leading to cracking and changes in the crystallinity of the coatings.

2.4 Obtaining Thin Layers by Thermal Spraying

Coating thermal spray materials has been known in the industry since 1910, introduced by Dr. Schoop (Switzerland).

The method consisted of projecting a jet of molten particles of material on the surfaces. The aim is to make layers resistant to wear and tear, such as decorative coatings, coatings for magnetic protection, obtaining metal powders, and obtaining patterns and molds. Due to temperature increases, the method has the advantage of deposits of various substances and very low thermal stress on the substrate [9].

Thermal spraying is a field that is successfully used in surface engineering due to its versatility. A wide range of deposition materials are used: metallic, composite, ceramic, memory alloy forms, Fig. 2.3.

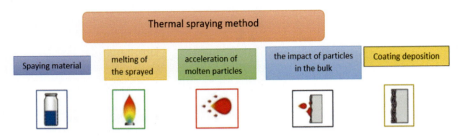

Fig. 2.3 Diagram of the thermal spraying procedure

The advantages and disadvantages of utilizing thermal spray are presented as the key steps in monitoring this field [10].

- High-temperature source – in this causes the method use electric arc, flame, detonation;
- Physical condition and material composition before deposition (granules or powder, wire or thin bar, pure substances, metal alloys, oxides, composite materials).

It has also been found that it is not easy to control the thickness and uniformity of thin layers obtained from compositions or other alloys by this method. To eliminate these shortcomings, the cathodic spray deposition method is used. This method has been widely used for a long time, but improvements in recent years have made it viable in making thin films with a very well-controlled structure.

The method consists of a luminescent discharge in an inert medium, Ar^+, and low pressure between two electrodes; the cathode is the source (the material to be deposited). The substrate is placed on the anode or in the space between the electrodes. The cathode destruction (erosion) phenomenon is used by bombarding it with ionized gas molecules in the discharge chamber. The dark space is critical when discharging near the cathode, where almost all the applied voltage falls.

The first method of forming layers by spraying consisted of entraining molten metal particles by a concentrated jet of air Fig. 2.4. In general, the variant was applied in spraying efficiently fusible metals and others that require layers made of such metals. The method is unproductive and, therefore, with limited activity. Currently, there is a wide variety of spraying techniques and installations.

One of the particular variants of flame spraying uses the energy of detonation of the oxygen mixture – C_2H_2. This form of spraying allows used the realization of layers by fusible materials. The source is the flame formed from the burned mixture of oxygen/fuel gas. The material is used in three forms; wire, bar, powder. Figure 2.5 shows the operating principle of the installations with oxygen gas flame using wire.

The material from the specified form wire, bar, or powder is fed through the central hole of the burner and melted by the flame. The jet of compressed air pulverizes the molten material into small fragments deposited on the bulk surface. The supply of material is produced with constant speed, generated by the rotational movement of some roles set in motion by a turbine that works with compressed air

2.4 Obtaining Thin Layers by Thermal Spraying

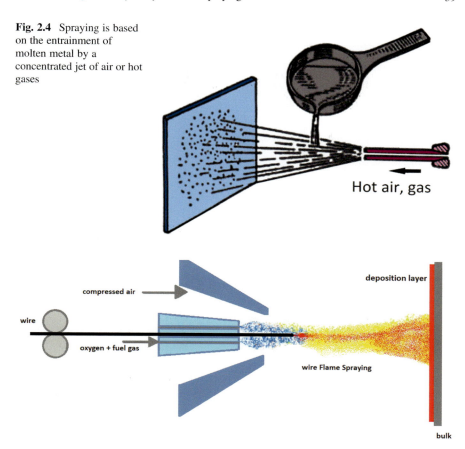

Fig. 2.4 Spraying is based on the entrainment of molten metal by a concentrated jet of air or hot gases

Fig. 2.5 Diagram of the installation oxygen gas flame

(also used for spraying) or with the help of an electric motor. In the vast majority of cases, the fuel gas used is acetylene. Propane, hydrogen, or methyl acetylene propane can also be used. Due to compressed air in the flame, some of it has a strong oxidizing character. The flame temperature does not exceed 2850 °C. Materials that have higher melting temperatures cannot use this method. High-performance metallic materials can be coated using technologies specific to new products [11].

2.4.1 Detonation Gun Spraying

This process is a "thermal spray process variation. It was introduced in the early 1950s and developed in Russia by Gfeller and Baiker working for Union Carbide. The controlled explosion of a fuel mixture gas, oxygen, and the powdered coating

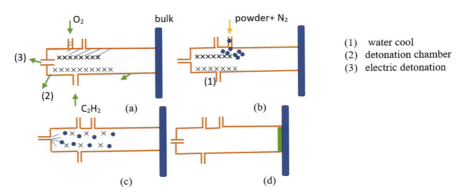

Fig. 2.6 Diagram of detonation; (**a**) filling the working chamber with detonating mixture; (**b**) powder supply; (**c**) detonation of the fuel mixture and acceleration of the powder particles; (**d**) deposition layer

material is used to melt and propel the material to the work piece" [12]. Oxygen and acetylene are fed in a tube closed at one end (Fig. 2.6).

The working chamber (2) presents with water coolers (1) diameter of 25 mm is powered by a strictly determined ratio. The powder is in the working chamber by using a stream of neutral gas, usually nitrogen. The gaseous mixture in which the powdered material is in suspension is ignited by means of an electric spark. As a result of the explosion and the release of heat, a shock wave is formed, which heats up and accelerates the dust particles in the support direction.

Detonation spraying is mainly intended to obtain hard and wear-resistant carbide layers with low proportions of binder metals, various oxides, or mixtures thereof. In practice, the layer thickness is usually chosen between 0.25 and 0.30 mm. The layer obtained by detonation is characterized by high density and adhesion value with support. During the detonation spraying, the substrate temperature remains low, not exceeding 200 °C. Basically, support does not deform and does not support any other physical changes.

The disadvantage of detonation spraying is dictated by the high noise level involved (approximately 140 dB). Another shortcoming of the process is the high prices of the installation and accessories [13].

2.4.2 Electric Arc Spraying Process

Electric arc spraying is the most widely used thermal spraying process (Fig. 2.7) and consists of:

- Melting the filler material with the electric arc formed between two wires connected to the poles of a welding rectifier;

2.4 Obtaining Thin Layers by Thermal Spraying

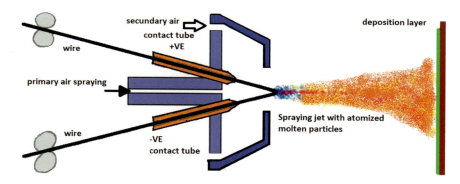

Fig. 2.7 Diagram of electric arc spraying

- Spraying the melt into very fine particles and transporting them to the deposition surface, which is performed with the help of compressed air [13, 14].

The advantage of the electric arc spraying method is the high productivity of the process when making high-performance materials. At a high current intensity value, for example, 750 A, the productivity for alloy steel is 36 kg/h, a value that several times exceeds the productivity of the oxygen flame gas spraying method. Compared to oxygen flame spraying, incomparably higher adhesion layers can be obtained on different materials. A disadvantage of the process is the overheating and oxidation of the sprayed materials.

2.4.3 Plasma Spraying

Plasma spraying is one of the thermal spraying processes that utilize a high energy heat source to melt and to accelerate fine particles onto a prepared surface [14]. Under industrial conditions, two types of burners are used to obtain plasma: plasma arc and plasma jet (Fig. 2.8). The layers obtained are characterized by high densities and good adhesion to the base material, thus forming new high-performance materials [15].

New materials with high structural and mechanical characteristics are produced through Plasma spraying by the interaction between the sprayed material and an active gas formed from a combination of argon/nitrogen or argon/nitrogen/hydrogen. These deposits are based on the diffusion of nitrogen into the base material in a vacuum-controlled atmosphere. Unlike other nitriding processes, such as PVD or CVD, this process allows a high deposition rate. Thick nitride layers solve the problems of corrosion and wear resistance of metallic and nonmetallic materials. These layers can be obtained by the reactive plasma spraying process, using either a direct current installation, a radio frequency installation, or a laser installation as a melting source.

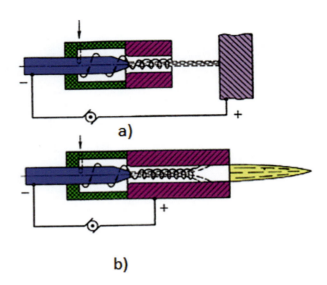

Fig. 2.8 (a) Plasma arc (b) Plasma jet

Flame spray deposition, electric arc spray deposition, plasma spray deposition, and detonation gun spraying deposition have the advantage of high productivity, using the application of successive and different layers, accessible only from some parts of the part. Efficiency is evaluated by the spray efficiency S, representing the average number of atoms emitted by the surface of the corpus bombarded by single ion. Using Boltzmann's equation in transfer theory, [16, 17] the following expression for the spray coefficient is obtained, a relation valid if the energy of the bombing ions does not exceed 1 keV:

$$S(E_i) = \frac{3}{4\pi^2} \alpha \cdot \gamma \cdot \frac{E_i}{U_0} \qquad (2.1)$$

where U_0 – is the energy of the surface bonds of the sputtering atoms. In the case of single charge ion bombardment, M_p is the mass of the amount of material sprayed in the unit of time on the unit area of the target, which can be calculated by the relation:

$$M_p = J_i \cdot S(E_i) \cdot \frac{A}{N_A} \cdot e \qquad (2.2)$$

where J_i – is the density of the ionic current at the spray surface, A – the atomic mass of the sprayed material, N_A – Avogadro's number, and e – the electron charge.

All deposition methods in this process aim to improve the surface properties of the substrate, especially the physical–mechanical and chemical ones.

2.5 Electrochemical Methods

Electrochemical deposits are made by electrolysis in aqueous solutions simple or complex containing the ion to be deposited. It is important for the working conditions that the processing is not accompanied by hydrogen release, which causes a decrease in the yield of the method. The pH value increase in the cathodic film, the presence of inclusions in the cathodic deposits, and the appearance of some internal tensions are possible due to the hydrogen overvoltage (U_H). The relation obtained empirically by Tafel[1] has the following form:

$$U_H = a + b \times \lg J \tag{2.3}$$

a – is a constant, the value is dependent on the nature of the cathode's material, b – is a constant, 0.12 V for aqueous solutions, J – is the current density. The thickness d of the layer deposited from a material characterized by the specific volume V_{sp} is given by the relation:

$$d = \eta \times V_{sp} \times J \times t \tag{2.4}$$

The relation gives the duration of the electrochemical process:

$$t = \frac{60 \times s \times \rho}{k \times \eta \times J} \tag{2.5}$$

s – is the deposition thickness; ρ – density of materials deposited; k – electrochemical equivalent; η – current efficiency.

The electrochemical method [18] is widely used in industry to protect the surfaces of parts, and depending on the metal used for coating, the process is called chrome plating, silvering.

In the study of electrochemical deposition methods, an important step is the precise control of parameters such as:

- The temperature at which the process takes place,
- The content of impurities,
- The pH value of the solution.

Anodic oxidation or anodizing is a widely used process for obtaining thin oxide layers of various metals. Usually, the thin layers obtained by anodizing have an amorphous structure, less often crystalline. The oxide layers are often subjected to the heat treatment that leads to their crystallization in polycrystalline or even monocrystalline layers.

---

[1] The Tafel equation is an equation in electrochemical kinetics relating the rate of an electrochemical reaction to the overpotential.

The disadvantage of obtaining thin layers by anodic oxidation is the possibility of soiling the layers with some impurities, which leads to defects and worsening of the dielectric properties of the film.

2.6 Chemical Vacuum Deposition Method (CVD)

The CVD technique has beendevised by Wöhler[2] in 1850 who deposited $Ti - C - N$, titanium carbonitride. The CVD at normal pressure consists of achieving the thin layer by the heterogeneous reaction on the surface of the substrate in a reactor where the total presence is 1 atm. The main advantages of chemical vapor deposition at normal pressure are the possibility of obtaining a variety of thin layers with electro-optical properties and perfectly determined chemical composition, on almost any substrate that can tolerate the deposition temperature, the deposition rates are relatively high, uniformity and reproducibility are very good, and deposition rates are constant over time.

CVD at low pressure has the advantage of obtaining uniform thicknesses on a very large number of pieces placed at close distances. Still, compared to CVD at normal pressure, the deposition rate is lower, so the process is limited to obtaining thicknesses of films a maximum of 1 μm.

Plasma chemical vapor deposition is the method that involves thin layers using an electric discharge, usually in radio frequency – RF [19]. Thin layer formation reactions take place mainly by the action of luminescent discharge. The advantage of the method is to obtain thin layers at low temperatures, including room temperature.

Some disadvantages of the method:

- The equipment can be complex in some cases;
- The variables that can be controlled are numerous (gas flows, temperature, etc.) [20];
- Vapors react in some cases with the substrate, or with the wall of installation, or even with the thin layer, which leads to the impurity of the growing layer;
- Thermodynamics and kinetic reactions involved in the deposition are very complex;
- The deposition processes involved deposition temperatures much higher than those required in the case of physical vapor deposition, which makes it inapplicable in various instances (substrates with relatively low melting point);
- In some cases, the necessary reactive sources are challenging to obtain or store [21];
- The toxic explosive or corrosive nature of the reactive gases used in the deposited process implies taking special precautions in their handling;

---

[2]Friedrich Wöhler (31 July 1800 – 23 September 1882) was a German chemist.

- Uniformity of deposition is often difficult to control;
- Masking certain parts is complex;
- The reaction by secondary products can be deposited in unwanted places, and removal is difficult.

References

1. A. D. Gloker, S. I. Shah, W.D. Westwood, Handbook of thin Film Process technology, Bristol, UK, Philadelphia: Institute of Physics 1995, pg. 214–315, (1995).
2. V. N. Antonov, J. S. Palmer, P. S. Waggoner, A. S. Bhatti, J. H. Weaver, *Nanoparticle diffusion on desorbing solids: The role of elementary excitations in buffer-layer-assisted growth,* Physical review b 70, 045406 (2004).
3. E. J. Ayers, T. Kujofsa, P. Rago, J. E. Raphael, *Heteroepitaxy of Semiconductors,* CRC Press, pg. 102–125, October (2016).
4. H. J.Y Chen, D. L. Yang, T. W. Huang, I. S. Yu, *Formation and Temperature Effect of InN Nanodots by PA-MBE via Droplet Epitaxy Technique,* Nanoscale Research Letters volume 11, 241, (2016).
5. O. Hellman, P. Steneteg, I. A. Abrikosov, S. I. Simak, *Temperature dependent effective potential method for accurate free energy calculations of solids,* Phys. Rev. B 87, 104111, (2013).
6. B. V. Shanabrook, J. R. Waterman, J. L. Davis, and R. J. Wagner, *Large temperature changes induced by molecular beam epitaxial growth on radiatively heated substrates,* Appl. Phys. Lett. 61, 2338, (1992).
7. L. Floroian, D. Floroian, M. Badea, *Advanced methods for thin layers of biomaterials obtaining with applications in implantology,* J.M.B. nr. 1, (2016).
8. W. S. Shi, H. Y. Peng, L. Xu, N. Wang, Y. H. Tang, S. T. Le, *Coaxial three-layer nanocables synthesized by combining Laser Ablation and Thermal Evaporation,* Advance Material, Volume12, Issue24, December, 2000, pg. 1927–1930, (2000).
9. N. Espallargas, *Future Development of Thermal Spray Coatings,* Woodhead Publishing Series in Metals and Surface Engineering: Number 65, Elesevier, pg. 17–21, (2015).
10. K. Simunovic, *Welding engineering and technology- Thermal Spraying,* pg. 2–11, EOLSS, 2010.
11. Popescu M. et al, *Acoperiri Termice și Recondiționări. Teme experimentale,* Politehnica Timisoara Publishing House, (2008).
12. F. J. Hermanek, *Thermal spray terminology and company origins,* (2001).
13. I. Fagoaga, G. Barykin, J. De Juan, T. Soroa, C. Vaquero, *The high frequency pulse detonation (HFPD) spray process* in: Thermal Spray 1999: United Thermal Spray Conference (DVS-ASM) Coatings, A., pp. 282–287, (1999).
14. L. Pawlowski, The Science and Engineering of Thermal Spray Coatings pg. 28–53, January 1, 1849.
15. M. Wang, *Composite coatings for implants and tissue engineering scaffolds,* Woodhead Publishing Series in Biomaterials, Pgs. 127–177, (2010).
16. S. Sigmund, E. Akgun, J. Meyer, J. Hubbuch, M. Worner, G. Kasper, *Defined polymer shells on nanoparticles via a continuous aerosol-based process,* Journal of Nanoparticle Research, 16(8), 2014.
17. S. Sigmund, M. Z. Yu, J. Meyer, G. Kasper, *An aerosol-based process for electrostatic coating of particle surfaces with nanoparticles,* Aerosol Science and Technology, 48(2), 142–149, 2014.

18. M. L. Sartorelli, A. Q. Schervenski, R. G. Delatorre, P. Klauss, A. M. Maliska, A. A. Pasa, *Cu–Ni Thin Films Electrodeposited on Si: Composition and Current Efficiency,* Physica status solidi (a) 187, No. 1, 91–95 (2001).
19. M. Vopsaroiu, M. J. Thwaitesa, G. V. Fernandez, S. Lepadatu, K. O'grady, *Grain size effects in metallic thin films prepared using a new sputtering technology*, Journal of Optoelectronics and Advanced Materials Vol. 7, No. 5, October 2005, pgs. 2713–2720, (2005).
20. L. Ji, Q. Ji, Y. Chen, X. Jiang, and K-N. Leung, *Sputter deposition of metallic thin film and direct, Presentation in 48th International conference on electron, ion, photon beam technology and nanofabrication*, June 1–4, 2004, San Diego, USA.
21. B.J. Lee, D. C. Lee, C. S. Kim, *Electrical properties of sputtered Ni-Cr-Al-Cu thin film resistors with Ni and Cr contents,* Journal of the Korean Physical Society, Vol. 40, No. 2, February 2002, pgs. 339–343, (2002).

Chapter 3
Vacuum Deposition

Abstract Modern installations for obtaining thin layers by evaporation and vacuum condensation are presented with automation and complexity. Some types of evaporation devices are mentioned for evaporation material used in thin film deposition. It is mentioned some types of evaporation devices for evaporation material used in thin film deposition. Thickness measurement by electrical and ultrasonic thickness measurement methods is revealed.

Keywords Evaporation chamber · Vacuum pumps · Vacuum valves · Electron beam · Ion beam · Laser beam heating · Evaporation devices · Thin conductor · Crucible · Thermal screen · Material for evaporation · Thickness measurement · Capacitive method · Inductive method · Resonant quartz method · Gravimetric method

Modern installations for obtaining thin layers by evaporation and vacuum condensation are presented with automation and complexity. Some types of evaporation devices are mentioned for evaporation material used in thin film deposition. The following discussion will introduce various evaporation tools for thin film deposition in evaporation materials. Thickness measurement by electrical and ultrasonic thickness measurement methods is revealed.

3.1 Vacuum Thin Film Deposition Installations

In general, an installation for obtaining thin layers by vacuum evaporation contains the vacuum system, pumps and traps, the evaporation chambers, and the electrical system. The vacuum system includes preliminary vacuum pumps, high vacuum pumps, connecting pipes, vacuum valves, liquid nitrogen (or helium) traps, vacuum measuring systems, etc.

Evaporation chambers, particularly in laboratory setups, are typically constructed using cylindrical bells composed of metal or glass, or cylinders crafted from materials such as glass or molybdenum. The electrical part of the installation

© The Author(s), under exclusive license to Springer Nature Switzerland AG 2023
N. Nedelcu, *Thin Films*, https://doi.org/10.1007/978-3-031-06616-0_3

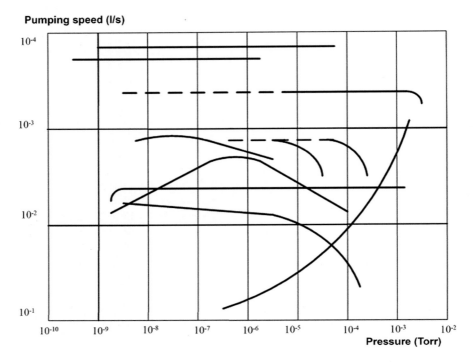

Fig. 3.1 Variation of pumping speed – pressure for vacuum pumps

includes the power supply and control system [1]. For the supply of mechanical pump motors, diffusion pump, and for obtaining currents in the evaporation process it is required to use a fairly high installed power (1 ÷ 10 kW). In the electric field, we can install a device for controlling the vacuum, film thicknesses, etc.

To obtain the high vacuum in the evaporation plant, quality vacuum installations are used, usually made by combining two or more pumping processes. In order to be able to choose a suitable combination of pumping procedures, a comparison of the different pumping systems will be used. It is assumed that the mechanical vacuum pumps were used to obtain a pressure of $10^{-2} \div 10^{-3}$ Torr. Oil diffusion pumps work in the range of $10^{-2} \div 5 \times 10^{-6}$ Torr; when using special oils DC 705 Silicone can provide a vacuum of 10^{-9} Torr without liquid nitrogen trap and $10^{-10} \div 10^{-11}$ Torr with liquid nitrogen traps.

Turbo-molecular pumps are less used, are expensive, and take up much space. Ghetto-ionic or magneto-ionic pumps are used in limited cases or combination with other pumping systems, work in small closed volumes, and can be easily saturated (the ghettos saturate). Figure 3.1 shows the variation of pumping speed – pressure for vacuum pumps. Absorption pumps reduce their pumping speed as the pressure in the enclosure decreases to $10^{-5} \div 10^{-6}$ Torr. In addition, absorption pumps, as magneto-ionic pumps, do not support the contact with atmospheric air, the situation is found in installations for obtaining thin layers; after introducing the air into the evaporation bell (for inserting the supports, the substance or removing the samples, etc.) it takes a very long pumping time (hours) to reach the initial vacuum. Silicone oil pumps have better-operating characteristics, are widely used in the thin film technique. The carbon compounds contained in the diffusion oil can be absorbed by titanium. For this purpose, a small titanium pump is placed in the working volume, which works simultaneously with the diffusion pump. In particular, titanium pumps work ionically and have a number of advantages (self-regulating, ultra-high vacuum, etc.). However, it is necessary that the magneto-ionic pump, including the titanium one, be introduced into the vacuum circuit after a pressure as limited as possible has been obtained in the enclosure. Compared to diffusion pumps, titanium pumps are at least twice expensive, which limits their use in current vacuum installations [2].

In Fig. 3.1 the cryogenic pumps have a constant pumping speed (between 10^{-4} and 10^{-9} Torr). The advantage of this type of pump is that it is not necessary to degas the enclosure beforehand. Combined pumping to achieve a high and ultra-high vacuum is currently used in modern installations for obtaining thin layers by thermal evaporation.

Most installations use the cryogenic system combined with another pumping procedure such as ghettos and diffusion pumps.

In obtaining high and ultra-high vacuum, the constructive and operational peculiarities of vacuum installation must also be taken into account. Some components in the installation utilize distinct elements, such as heated rubber gaskets, that can substantially reduce gas and modify their characteristics and flexibility. This is particularly notable when operating under high temperatures and low pressures.

3.2 Evaporation Devices

Evaporation devices are heating devices used in evaporation. The analysis of evaporation devices contributes to a more complete understanding of the technology and properties of thin films, which depend on the evaporator's nature, type, and construction. Evaporation devices can be with direct or indirect heating of the evaporation material. To evaporate materials through direct heating, evaporators can take the form of either a conductor or a strip extending between two contacts (as shown in Fig. 3.2a) or a set of filaments attached to metal supports (as seen in Fig. 3.2b) [3].

This method is used especially in the evaporation of metals Al, Ti, Cu, Ni, etc. Mainly due to the destruction of the evaporator and nonuniformity, evaporation is less used in the technique of obtaining thin layers.

In the case of indirect evaporation, special devices are used with which the substances are heated to the evaporation temperature. These devices are divided into three main groups: resistive, inductive, and electron or ion beam, depending on the heating mode. In resistive evaporators, the thermal energy is obtained due to the Joule effect produced when the current passes through the resistive device (band, thin conductor, tube, etc.). These devices are very common due to the simplicity of their construction, power supply, and handling. In Fig. 3.3, types of resistive evaporation devices are represented.

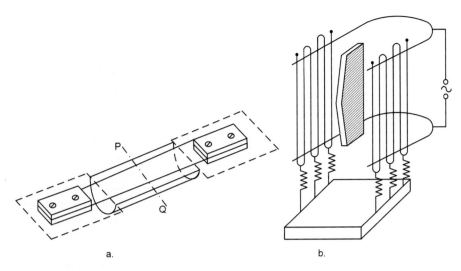

Fig. 3.2 Evaporator types with direct heating (**a**) conductor or a strip extending between two contacts (**b**) a set of filaments attached to metal supports

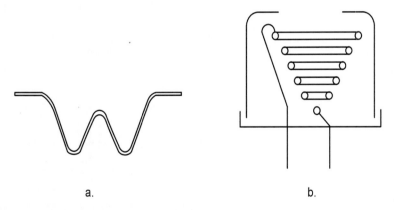

Fig. 3.3 Types of resistive evaporation devices (**a**) band or thin conductor (**b**) tube

Their shape and size, type of construction, and other characteristics are determined by the nature and substantial volume of the evaporator and, respectively, by the conditions that resistant evaporators must meet.

The main conditions imposed for resistive evaporators are mentioned:

- The material to be evaporated in the liquid state must make thermal contact with the material of the evaporator, have adhesion to its surface.
- Although it is necessary to have good adhesion between the evaporator and the material to be evaporated, there must be no chemical reactions between them.

3.2 Evaporation Devices

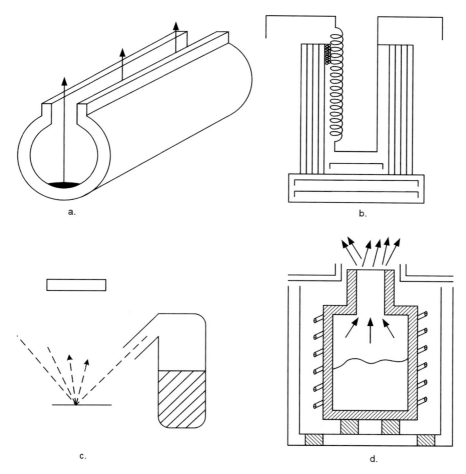

Fig. 3.4 Types of evaporators of high capacity and efficiency (**a**) Horizontal Tube (**b**) Vertical Tube (**c**) Long Tube Vertical (**d**) Forced Circulation

In the case of evaporators in the shape of V, U, W, filiform, etc., pieces of evaporating materials are used, placed on the evaporator, which will be heated by passing electric current. To prevent the waste of material and to avoid the substance's deposition on other surfaces, special screens or evaporators in the form of baskets or trays and strips are used (Fig. 3.4a–d).

Thermal screens increase the efficiency of the evaporator, avoiding heat loss. On the other part, for the evaporation of powdered substances (dielectric substances, semiconductors, etc.), it is necessary to use metal trays, quartz baskets with tungsten, or molybdenum heater.

Crucibles from aluminum (Al_2O_3), beryllium oxide (BeO), or thorium (ThO_2) are used to evaporate powder materials. In many cases, graphite crucibles are used, but they have the disadvantage of releasing a large amount of gas during evaporation. At

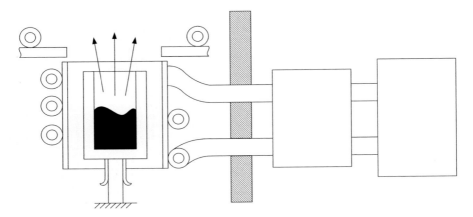

Fig. 3.5 Evaporation installation's scheme with inductive heating

high temperatures, metals with some exceptions (Ag, Be, Sr, etc.) react with carbon to form different carbides. Semiconductors also react with graphite at high temperatures. For example, evaporation of Si from graphite crucible can only be done with special devices. To prevent the expulsion of the evaporating substance from the crucible, special-purpose crucibles with a tubular roof in the form of a sieve are used (Fig. 3.4c).

In modern evaporative installations, especially the industrial ones, special devices are used for supplying the evaporator with the substance to be evaporated, which at the desired time interval lets the substance fall into the crucible; if the substance is found in the form of threads or thin strips, they are unwound on special spools or are gradually introduced into the evaporation region.

For the evaporation of larger quantities of substance and to eliminate the direct contact between the substance and the evaporator, evaporation installations with inductive heating are used. The functional diagram of such an installation is shown in Fig. 3.5. Usually, the inductor coil, consisting of 3–4 turns, is made of copper pipe through which the water passes for cooling. A generator current that works at 1–2 MHz is used. Similar plants are used in semiconductor technology to grow and purify single crystals.

The electron beam, ion beam, or laser beam heating method is a modern and particularly effective method for evaporating materials. Figure 3.6 shows the working diagram with the laser beam. A pulsed laser (1) such as carbon dioxide is used. The lens (2) focuses the beam so that it passes easily through the window (4) and focuses on the substance (3) to be evaporated. Many hard-fusible materials such as W, Mo, Ti, etc., can be evaporated.

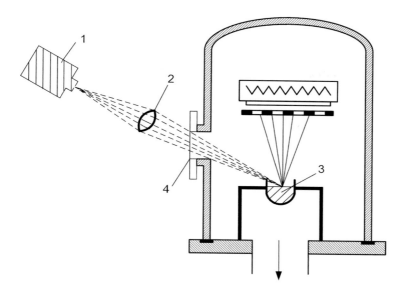

Fig. 3.6 Evaporation installation's scheme through laser heating

3.3 Important Materials for Evaporation

Oxides are essential materials for evaporation, generally are used for abrasion resistance, and have a variety of refractive indexes which can be combined with other materials to obtain a high transmission.

SiO_2 – can be deposited by electron beam evaporation with IAD [4] or without IAD, electron beam evaporator [5], equipped with copper or molybdenum crucibles, by ion beam, by reactive, RF sputtering. It is used especially for VIS-NIR spectral range.

SiO – is also deposited by electron beam evaporator, or equivalent, equipped with graphite crucibles. This kind of material is used in particular for infrared spectral range [6].

TiO_2 – it is deposited using an electron beam evaporator and equipped with copper or molybdenum crucibles. TiO_2 and SiO_2 are the two materials that can be optimized using optical calculation software for various coatings in the VIS-NIR [7].

ZrO_2 – can be deposited using an electron beam evaporator equipped with graphite crucibles. This material is used in particular for visibile spectral range [8].

Y_2O_3 – using electron beam evaporator, equipped with graphite crucibles, [9] can be combined with materials for infrared spectral coating.

Ta_2O_5 – evaporates from copper or molybdenum crucibles.

Ag – can be evaporated by a resistant tungsten evaporator or. graphite crucibles.

Al – is evaporated from copper or molybdenum crucibles or can be used as a tungsten wire evaporator. The evaporator can be in the form of two wires in electrical contact and are placed pieces of aluminum wire in the shape of V or U. The flat surface made of wolfram wires, rigidly locked at the ends, on which are set linear pieces of aluminum wire.

Cu – can be evaporated in a resistive evaporator made of tungsten, or electron beam evaporation needs a graphite crucible.

Ge – is a material used for infrared spectral range, can be evaporated in a graphite crucible.

MgF_2 – is widely used in industry especially for obtaining thin films in the visible spectral range, which can be evaporated using copper or molybdenum crucibles.

CaF_2, YF_3 – it is used in coating industry especially for obtaining thin films in the infrared spectral range, can be evaporated using copper or molybdenum crucibles.

ZnS – this material can be evaporated from a resistive evaporator by molybdenum sheet; it is used especially for the infrared area.

3.4 Thickness Measurement Methods

One of the simplest methods of measuring the thickness of layers is based on the dependence of the electrical resistance on the deposited layer thickness and the average length of the carriers in the layer. Last clarification is necessary, this parameter determines the mobility of the carriers, and implicitly the resistance of the layers depends very much on the crystalline structure and the conditions of sample preparation. The measurement of the thickness layers by resistance is applicable in the case of metals and semiconductor layers with a low resistance Fig. 3.7. Conditions of layer preparation can significantly influence the resistivity of

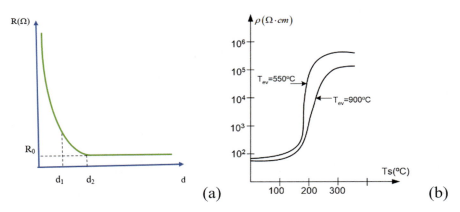

Fig. 3.7 Resistance variation of thin layers with the thickness (**a**) and resistivity with temperature (**b**)

3.4 Thickness Measurement Methods

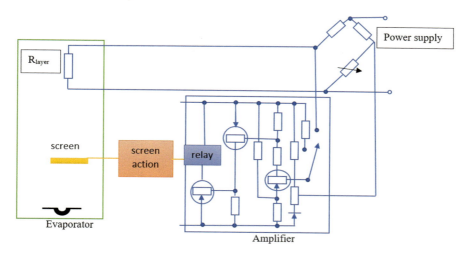

Fig. 3.8 Block diagram for controlling and measuring the layer thicknesses

the sample and it depends on the deviation of the composition from the stoichiometry, in the case of composite substances, the doping (impurity) of the material (especially in the case of semiconductors), of crystalline structure, of surface conditions, etc.

Depending on the stoichiometric composition's deviation, the layer's resistivity can vary, in some cases, by 5–6 orders of magnitude (Fig. 3.7b).

The electrical resistance can only measure and control layer thickness under strict experimental conditions. To accurately prepare samples and reproduce the electrical properties, methods for measuring thickness must be employed. The method is especially applicable to metals or other simple conductive materials after a necessary prior calibration of the device in series production conditions. The layer resistance is usually measured during deposition, using a power supply or other special devices. In Fig. 3.8, a block diagram is presented, intended to control the thickness during deposition time.

The thin (metallic) layer also acts as a control and steering element; when the layer thickness reaches the predetermined level, obviously characterized by a specific electrical resistance, points become unbalanced and the signal is strongly amplified, relay R actuates the screen E, and the evaporator contacts; in this way, evaporation is stopped when the required thickness is reached. The scheme can be designed in different ways; for example, the relay needs to activate the screen and contacts when the points are balanced [9]. The described system allows the control of the resistance of the layers in the range $10–10^3$ Ω, but the appropriate modification of the scheme can extend the limits.

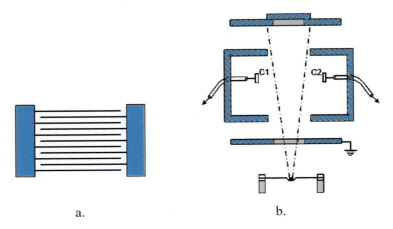

Fig. 3.9 Variants of the capacitive method for measuring the thickness of thin layer (**a**) Plate capacitor (**b**) Capacitive-based pressure sensor

3.4.1 Capacitive Method

The capacitive method was also used to control the thickness of thin layer deposition. Keister and Scapple [10] used support containing a "comb" type capacitor, obtained by photolithographic methods from aluminum foil (Fig. 3.9a).

The substance deposited between the "teeth" of the capacitor increases the capacity, its variation being highlighted by appropriate means (sensitive bridge methods). The method is especially suitable for thin dielectric layers, characterized by a high value of permittivity, but has the disadvantage of device inconvenience. The sample remaining fixed on the "capacitor" placed on the support; in addition, oxides, water vapor, etc., on the surface of the support can seriously influence the measurements.

Another variant of the capacitive method for measuring thickness layer deposition is illustrated, after Baker-Jarvis [11], in Fig. 3.9b. The method consists of measuring the variation (increase) of the capacity of the plate system C_1–C_2 when the particle beam of the evaporated substance passes through the space between them. Obviously, the measurements are reproducible for evaporation conditions (evaporation temperature, bulk temperature, pressure, evaporation duration, material composition, etc.) rigorously constant. However, this method is not precise enough, so it is very rarely used.

3.4.2 Inductive Method

A very interesting electrical method for measuring the thickness layers, especially if the bulk is made of ferromagnetic materials, is the inductive method. The principle of the method is illustrated in Fig. 3.10. It uses a "transducer" in the form of a metal

3.4 Thickness Measurement Methods

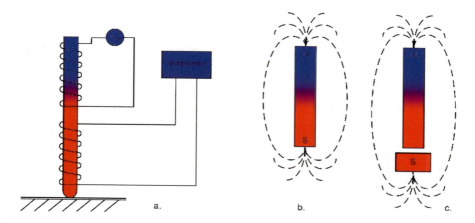

Fig. 3.10 The inductive method of measuring the thickness thin layer (**a**) Probe according to the magnetic inductive method (**b**) Penetration depth of the eddy current on a flat surface (**c**) Penetration depth of the eddy current on a concave surface

or ferrite rod, with the bottom rounded so as not to damage the layer whose thickness is measured. Two coils are mounted on the rod. Through coil 1 passes an alternating current which induces in coil 2 an alternating current of the same frequency, measured with a voltmeter (sometimes with a numeric voltmeter).

3.4.3 Resonant Quartz Method

This method is widely utilized for its accuracy and convenience. The procedure is based on the variation of the quartz frequency oscillator (increase) of the quartz plate's mass (and thickness). From this point of view, the method can be included in the category of microbalances for measuring deposited masses. The method's main advantage consists in its universality, being applicable also in the case of deposited metal, semiconductors, dielectrics, etc.

The working scheme is usually complex; the device also plays the monitor-controller on evaporation and stops the process when it reaches the desired thickness. To calculate the thickness layer d, it is required to know the density γ of the material considered a thin layer [12]:

If Δm it is the mass of the deposited layer, $V = \frac{\Delta m}{\gamma}$, the thickness $d = \frac{\Delta m}{\gamma S}$, results, S is the surface of the quartz crystal on which the thin layer was deposited. The frequency oscillation of the quartz crystal varies with Δm from the initial value m_o of mass.

The selection of the operating frequency (f) depends on the thickness range that requires measurement [13]. A high frequency must be chosen to achieve high sensitivity when measuring very thin layers. Using complex and sensitive electronic equipment, variation of frequency at ~20 Hz (at $f = 10^7$ Hz) can be highlighted

58

which allows variations measure of the mass order of 10^{-8} $\mu g/cm^2$ and respectively thickness between 100 and 100,000 Å.

3.4.4 Gravimetric Method

One of the oldest methods of measuring the thickness layers is the gravimetric method. The method consists of measuring the weight of the substrate (with an analytical balance) before and after deposition. The accuracy of the method depends on:

- The error of measuring the weight deposited layer is given by the error of the balance and by the error introduced by the modification of the weight of the layer by absorption in gases when releasing into the atmosphere;
- Measurement error of the substrate area.

References

1. A. Katsageorgiou, *Vacuum diffusion pumps: description of geometry, operation principals, design characteristics, specifications and simulation approaches*, Diploma Thesis, 2017.
2. J. Vickerman, I. Gilmore, *Surface Analysis – The principal techniques*, 2-nd ed., Wiley, 2009.
3. A. O. Adeyeye, G. Shimon, *Chapter 1 - Growth and Characterization of Magnetic Thin Film and Nanostructures*, Handbook of Surface Science, Volume 5, 2015, Pages 1–41.
4. M. Alvisi, G. De Nunzio, M. Di Giulio, M. C. Ferrara, M. R. Perrone, L. Protopapa, and L. Vasanelli, *Deposition of SiO$_2$ films with high laser damage thresholds by ion-assisted electron-beam evaporation*, Applied Optics Vol. 38, Issue 7, pp. 1237–1243 (1999).
5. G. Honciuc, N.Nedelcu, A. Zorilă, L.Rusen, L. Neagu, I. Dumitrache, A. Stratan - *Antireflection coatings for 650 nm and 1064 nm with high laser induced damage threshold*, Mini - Symposium / Workshop on Laser – Induced Damage and Laser Beam Characterization –, May 20–24th, 2013.
6. A. Hjortsberg, C G Granqvist, *Infrared optical properties of silicon monoxide films*, Appl. Opt. 19(10):1694–6, 15 May, (1980).
7. H. Selhofer, E. Ritter, and R. Linsbod, *Properties of titanium dioxide films prepared by reactive electron-beam evaporation from various starting materials*, Applied Optics Vol. 41, Issue 4, pp. 756–762 (2002).
8. R. E. Klinger and C. K. Carniglia, *Optical and crystalline inhomogeneity in evaporated zirconia films*, Applied Optics, Vol. 24, Issue 19, pp. 3184–3187 (1985).
9. A. Kasikov, *Optical inhomogeneity model for evaporated Y$_2$O$_3$ obtained from physical thickness measurement*, Applied Surface Science, Vol. 254, Issue 12, 15 April 2008, Pgs. 3677–3680, (2008).

References

10. F. Z. Keister and R. Y. Scapple, *A Thin-Film Multilayering Technique for Hybrid Microcircuits*, Solid State Technology, vol. 17, no. 5, pp. 44–47, May 1974.
11. J. Baker-Jarvis, C. Jones, B. Riddle, M. Janezic, R. G. Geyer, J. H. Grosvenor, and C. M. Weil, *Dielectric and magnetic measurements: a survey of nondestructive, quasinondestructive, and process-control techniques*", Research in Nondestructive Evaluation, Vol. 7, pp. 117–136, 1995.
12. A. W. Czanderna C. Lu, Chapter 1 - *Introduction, History, and Overview of Applications of Piezoelectric Quartz Crystal Microbalances*, Vol. 7, 1984, Pages 1–18.
13. L. Chih-Shun, *Investigation of film-thickness determination by oscillating quartz resonators with large mass load*, Journal of Applied Physics 43, 4385 (1972).

Chapter 4
Morpho-structural Characterization

Abstract The characterization methods studied the physical and chemical parameters of new material to ensure the quality of the material are presented. This information is very important and is used by many engineers in the coating industry. In solid-state physics, information of the structural material nature is obtained by diffraction techniques (ND-diffraction, XRD-diffraction). This study aims to identify the structural, morphological, and crystalline properties required with the help of microscopy (AFM, SEM, TEM microscopy) and X-ray techniques.

Keywords Characterization methods · Structural and morphological analysis · Neutron diffraction · X-ray diffraction · Electron diffraction · Microscopy techniques · Atomic force microscopy (AFM) · Scanning electron microscopy (SEM) · Transmission electron microscopy (TEM) · X-ray spectroscopy · Electron backscatter diffraction (EBSD) · Auger electrons · Scattered electrons · Kikuchi lines · High-resolution transmission electron microscopy (HR-TEM) · Selected area electrons diffraction (SAED)

The characterization methods studied the physical and chemical parameters of new material to ensure the quality of the material are presented. This information is very important and is used by many engineers in the coating industry. In solid-state physics, information of the structural material nature is obtained by diffraction techniques (ND-diffraction, XRD-diffraction). This study aims to identify the structural, morphological, and crystalline properties required with the help of microscopy (AFM, SEM, TEM microscopy) and X-ray techniques.

4.1 Methods for Characterizing Thin Layers

To ensure the quality of vacuum-deposited layers, it is essential to identify and evaluate the physical and chemical parameters that define them. These parameters can be grouped as follows:

© The Author(s), under exclusive license to Springer Nature Switzerland AG 2023
N. Nedelcu, *Thin Films*, https://doi.org/10.1007/978-3-031-06616-0_4

- Morphological parameters (appearance, topography, structure, thickness);
- Mechanical parameters (density, coefficient of friction, adhesion, hardness, elasticity);
- Chemical parameters (density, stability, toxicity);
- Electrical parameters (electrical conductivity, permeability, polarization, dielectric constant, saturation induction for magnetic films);
- Technical parameters (melting temperature, coefficient of thermal expansion, thermal conductivity, vapor pressure).

4.2 Structural and Morphological Analysis

In solid-state physics, information of the structural material nature is obtained by diffraction techniques using various radiations with different wavelengths such as:

- Neutron diffraction,
- X-ray diffraction,
- Electron diffraction.

From the point of view of mathematical-physical formalism, all these methods are identical. In the case of a perfect crystal, the plane waves reflected on various atomic planes interfere. The beam intensity will be maximum when the optical path difference is an integer number of wavelengths, Fig. 4.1.

$$D = 2d(h,k,l) \sin \theta \qquad (4.1)$$

where d – is the distance between atomic planes, θ – is the angle of incidence.

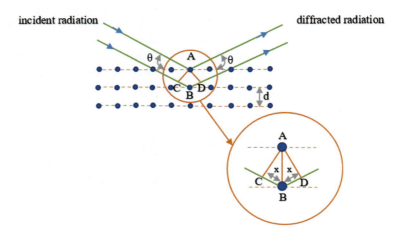

Fig. 4.1 The principle of X-ray diffraction

4.3 Neutron Diffraction (ND)

The X-ray range includes electromagnetic radiation with wavelengths between 0.2 and 200 Å.

In the case of amorphous materials, the X-ray division amplitude for a single atom is given by the relation [1]:

$$A_i = f_i \exp\left(-i\vec{q} \cdot \vec{r}_i\right) \tag{4.2}$$

f_i – atomic scattering factor; $\vec{r}_i$ – position of the atom i.

The intensity of the scattered radiation $I_{eu}\left(\vec{q}\right)$ for i atoms is given by the relation:

$$I_{eu}\left(\vec{q}\right) = \left(\sum_i A_i\right)\left(\sum_j A_j^*\right) = \left(\sum_i f_i \exp\left(-i\vec{q}\right) \cdot \vec{r}_i\right)$$
$$\left(\sum_j f_j^* \exp\left(-i\vec{q}\right) \cdot \vec{r}_j\right) = \sum_i \sum_j f_i^* f_j \exp\left(i\vec{q}\right) \cdot \vec{r}_{ij}) \tag{4.3}$$

$\vec{r}_{ij}$ – the vector which unites m and n atoms.

The Debye equation [2] is:

$$I_{eu}(q) = \sum_i \sum_j f_j^* f_i \frac{\sin qr_{ij}}{qr_{ij}} \tag{4.4}$$

The structure factor $S(q)$ is defined as:

$$S(q) = \frac{I_{eu}(q)}{N <f(q)>^2} \tag{4.5}$$

N – the number of atoms.

The X-ray diffraction image of a material depends on the diffraction method and the material's structural characteristics. Among the methods of structural characterization of nanocomposites, the most common method is X-ray diffraction which is much more convenient than transmission electron microscopy (TEM). The preparation of samples is much easier.

4.3 Neutron Diffraction (ND)

Various experiments have been performed that have shown neutrons can also present the diffraction phenomenon, the method being applied to the study of solid-state structures since 1936. The application of neutron diffraction covered the problems of X-rays and electron-graphs being used to determine magnetic structures, taking advantage of the fact that the neutron is electrically neutral and has its magnetic moment.

The construction of nuclear reactors made it possible to obtain neutron beams, very similar to X-ray diffraction. In 1945, the USA built the first diffractometer, or "neutron spectrometer" [3].

In the solid state, neutrons or other particles with undulating properties are used to study the arrangement of atoms. Their speed requires that the wavelength be corresponding to the order of the distance between the atoms. The relation gives the wavelength:

$$\lambda = \frac{h}{p} \tag{4.6}$$

where λ – wavelength, h – Planck's constant, p – neutron pulse.

At the exit of the reactor, some neutrons undergo numerous collisions with atoms at temperature T, and their average square velocity corresponding to temperature T is defined by the relation:

$$\lambda^2 = \frac{h^2}{3mKT} \tag{4.7}$$

m – neutron mass, T – temperature, K – Boltzmann's constant.

The two relations show that the wavelengths corresponding to the temperature of 0 and 100 °C are 1.55 and 1.33 Å. In the solid study, these wavelengths are the most applied, the neutrons corresponding to these temperatures are the easiest to obtain. In the reactor, these neutrons are braked by numerous collisions with the atoms of the braking medium (graphite or heavy water), and at the reactor temperature, they tend towards thermal equilibrium.

The neutron velocity distribution in the reactor is of the Maxwellian type corresponding to a temperature of 100 °C. A collimator is in the reactor wall and can extract a neutron beam from the reactor. If $\nu\lambda d\lambda$ [3, 4] is the number of neutrons emitted in a second in the wavelength range λ and $\lambda + d\lambda$, then:

$$\nu_\lambda = \frac{2N_1}{\lambda} \left(\frac{E}{KT} \right)^2 \tag{4.8}$$

N_1 – is the total number of neutrons that are emitted in 1 s, E – neutron energy with wavelength λ.

Figure 4.2 shows the operating diagram of a neutron diffractometer: neutrons and X-rays interact with matter in different ways: neutrons interact with the nucleus; X-rays interact with electrons.

The angle of each element is adjusted so that Bragg's law selects the energy transfer and the specific pulses according to relation (4.1).

4.3 Neutron Diffraction (ND)

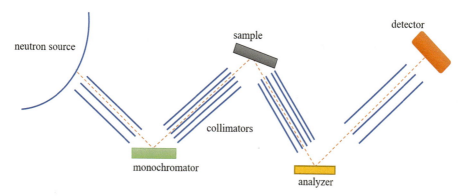

Fig. 4.2 Diagram of a neutron diffractometer

X-ray diffraction method and neutron diffraction booth methods can be correlated to determine the structural factors. The existing differences in the ND and XRD spectra are found in different [5] values of the weighting factors, (w_{ij}), from the composition of the partial structure factors, $S_{ij}(Q)$, defined as:

$$S(Q) = \sum_{i,j}^{k} w_{ij} S_{ij}(Q), \tag{4.9}$$

$$w_{ij}^{ND} = \frac{c_i c_j b_i b_j}{\left[\sum_{i,j}^{k} c_i b_j\right]^2} \tag{4.10}$$

$$w_{ij}^{XRD}(Q) = \frac{c_i c_j f_i(Q) f_j(Q)}{\left[\sum_{i,j}^{k} c_i f_i(Q)\right]^2}, \tag{4.11}$$

where c_i, c_j – are the molar fractions of the components, b_i, b_j – neutrons and $f_i(Q)$, $f_j(Q)$ – the scattering amplitudes of the X-rays, and k is the number of elements in the sample.

Experimental $S(Q)$ diffraction data can be simulated by the Reverse Monte Carlo (RMC) method, an efficient tool used to model disordered structures. From the RMC configuration, the following can be calculated:

- The partial distribution function of pairs of atoms,
- Coordination numbers,
- Bond-angle distribution functions.

Several RMC simulations can be performed by slightly changing the interatomic distances to different pairs of atoms. Thus, the experimental $S(Q)$ factors obtained by both methods (XRD and ND diffraction) can compare with the data obtained from the RMC modeling. Interatomic distances can be taken from previous studies or the literature based on similar or very appropriate compositions.

As a result, the two diffraction methods provide complementary information, and both types of measurement are needed to have a real structure for the investigated samples.

4.4 Microscopy Techniques

In simple terms, the microscopy provides magnified images of objects whose resolution cannot be compared to that of the human eye (100 μm). The image provided is obtained according to the technique provided. The optical microscope, for example, is based on visible radiation as a signal source. Other microscopy techniques are based on various signal sources such as electrons. Electron microscopy is a crucial technique in the study of structures and morphology material.

The most widely used is transmission electron microscopy, which uses either a direct or an indirect process and allows obtaining images of crystalline networks, with a resolution below 10 Å, which favors the highlighting of crystalline defects. The number of defects observed is dictated by the high contrast provided by the imagination, the electron diffraction, and the image's magnification over 20,000 times.

The direct process is applied on thin layers without support (obtained by chemical jet corrosion), with approximate thicknesses 100 Å ... 5000 Å, and is based on obtaining contrast pictures, both due to the differential absorption of electrons in various sample regions, especially the effects of electronic diffusion and interference.

The indirect process is used to obtain pictures of films deposited on different bulk samples. The resolution power of the transmission electron microscopy is 2 Å, being surpassed by any other method.

4.4.1 Atomic Force Microscopy (AFM)

An atomic force microscope creates images by monitoring the motion of an analyzer tip that is scanned along a surface. The device allows direct viewing of objects of the order of nanometers [6].

The atomic force microscope is used to study phenomena such as adhesion, corrosion, friction, lubrication, plating, and polishing. Using AFM measurements, you can view the surface at the nanoscale. The roughness of the analyzed surfaces is a parameter of interest. The atomic force microscope is part of the family of scanning microscopes. Various scanning microscopy techniques have become important in recent years in the study of thin films. Scanning microscopes are used to study the local properties of materials from the atomic to the micron level. All involve finely

4.4 Microscopy Techniques

Fig. 4.3 Schematic of an atomic force microscope

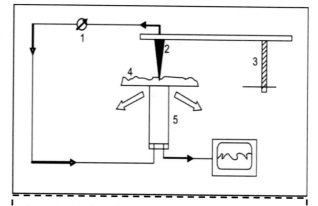

1. A device for detecting the vertical position of the analyzer tip;
2. The tip contacts the sample surface;
3. Slider for positioning the tip analyzer near the sample;
4. Sample;
5. Piezoelectric scanning device that moves the sample.

scanning the sample's surface, using an analyzer tip placed near the sample, turning the information into an image of the surface [7–9]. Figure 4.3 is the general scheme of an atomic force microscope.

The atomic force microscope examines the surface of a sample with a very sharp analyzer tip, several microns long and often less than 100 in diameter. The tip of the analyzer is fixed to the free end of a cantilever with a length of 100–200 μm. The forces between the analyzer tip and the sample surface move the cantilever. A detector measures the displacement of the cantilever as the analyzer tip scans the sample. Measuring cantilever movements allows a computer to generate a map of the surface topography. Several forces contribute to the movement of the cantilever. The force most often associated with atomic force microscopy is the interatomic force, van der Waals. There are three primary modes of operation of the atomic force microscope (AFM): AFM contact mode, AFM noncontact mode, AFM intermittent contact mode.

All these modes are based on recording the cantilever's oscillation near or in contact with the analyzed sample. The interaction of various forces between the sample and the analyzer tip causes it to flex. The dependence of this distance force between the analyzer tip and the sample is shown in Fig. 4.4.

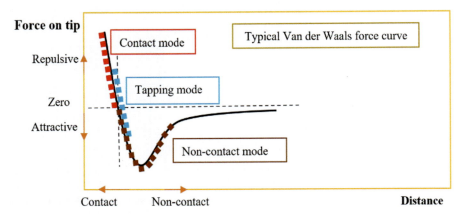

Fig. 4.4 The action of the interatomic force as a function of distance from the sample

The amplitude parameters that provide roughness and texture information can be calculated from the following relationships:

- The mean square roughness (RMS), S_q, is defined as:

$$S_q = \sqrt{\frac{1}{AB} \sum_{k=0}^{A-1} \sum_{l=0}^{B-1} [z(x_k, y_l) - \mu]^2} \qquad (4.12)$$

μ – is the average height, defined as:

$$\mu = \frac{1}{AB} \sum_{k=0}^{A-1} \sum_{l=0}^{B-1} z(x_k, y_l) \qquad (4.13)$$

- The asymmetry surface degree (skewness – S_{sk}) is defined by the formula:

$$S_{sk} = \frac{1}{ABS_q^3} \sum_{k=0}^{A-1} \sum_{l=0}^{B-1} [z(x_k, y_l) - \mu]^3 \qquad (4.14)$$

- The distribution probability coefficient (kurtosis – S_{ku}):

$$S_{ku} = \frac{1}{ABS_q^4} \sum_{k=0}^{A-1} \sum_{l=0}^{B-1} [z(x_k, y_l) - \mu]^4 \qquad (4.15)$$

- The height "peak to peak," S_y, defined by: $S_y = z_{\max} - z_{\min}$ (z is the normal axis to the plane (x, y) of the surface) – which can be understood as the thickness of the roughness layer.

4.4 Microscopy Techniques

Compared to other microscopy techniques, AFM is distinguished by the simplicity of sample preparation. Although the AFM can theoretically [10, 11] view a specimen of any size, the measurements are still limited by:

- Sample size,
- Scan speed,
- The memory required for electronic data storage,
- The maximum measuring distance of the scanner,
- The substrate used for fixing the sample.

The AFM was invented in 1986 by Binning, Quate. The main advantage is that it is a nondestructive technique, which can visualize the microscopic scale of the generally unstable organic molecule when interacting with electronic beams. AFM has special applications in surface science (micro- and nano-topography) [11].

4.4.2 Scanning Electron Microscopy (SEM)

The development of electron microscopy, since 1931, focused on two types of electron microscopes:

- Transmission electron microscopy (TEM);
- Scanning electron microscopy (SEM).

Due to its ease of use and the multitude of information it can give, scanning electron microscopy is often preferred as a technique for analytical microscopy. In scanning electron microscopy, a stream of high-energy electrons explores the surface of a material. The flow of electrons interacts with the material, producing a variety of signals – secondary electrons, backscattered electrons, X-rays, photons, etc. – each of which can be used to characterize a material with specific properties [12].

Two sets of information about the studied material can be obtained by scanning electron microscopy:

- *Topographic image*: scanning the surfaces of the material with a strong beam focused by energetic electrons – topographic images can be obtained that can reach 2 nm resolutions [13].
- *Micro-characterization*: by scanning electron microscopy, a series of micro-characterizations of the studied material can be obtained:

    (a) *The analysis composition (EDX)* – using X-ray spectrometry with dispersive energies, SEM can give the chemical spectrum and spatial distribution of the chemical elements specific to the material studied on a submicron scale;

    (b) *Luminescence analysis* – wave light scattering and/or cathodoluminescence can be used to study the identity, efficient recombination, and defect distribution of materials;

70   4   Morpho-structural Characterization

Table 4.1 Comparison between examination of optical microscope and scanning electron microscope

Sample (characteristics)	Optical microscope	Scanning electron microscope (SEM)
Examination mode	Atmospheric pressure	Vacuum
State of aggregation	Solid liquid	Solid
Electrical conductivity	It is not necessary	It is necessary in vacuum
Depth's field	Small	Large
Maximum resolution	3200 Å (usually) 1000 Å (special condition)	35 Å (usually) 5 Å (field emission)

Table 4.2 Comparison between scanning electron microscopes (SEM) and transmission microscopes (TEM) characteristics

Characteristics	Scanning electron microscope (SEM)	Transmission electron microscope (TEM)
Maximum resolution	5 Å	1 Å
Acceleration voltage	0.2 ÷ 50 kV	0.2 ÷ 1250 kV
Sample form	Solid	Thin film
Pictures	Surface topography; allows stereoscopic pictures	Internal structure; high-resolution picture

(c) *Structural micro-characterization* – with this technique the specific type of crystallinity, orientation, and quality of individual crystallites in polycrystalline materials can be studied [12–14].

Tables 4.1 and 4.2 compare the important features of optical microscopes, scanning electron microscopes (SEM), and transmission electron microscopes (TEM) [14].

The difference between SEM and TEM is that SEM creates an image by detecting reflected electrons, while TEM uses transmitted electrons (electrons that are passing through the sample) to create an image. As a result, TEM offers valuable information on the inner structure of the sample, such as crystal structure, morphology, and stress state information. At the same time, SEM provides information on the sample's surface and its composition.

SEM uses an electron detector to convert the radiation of interest into an electrical signal for manipulation and presentation via the electronic signal processing system. Most SEMs are equipped with an Everhart – Thornley detector, which works as follows: scintillating material is hit by an energetic electron; this collision produces photons driven by total internal reflection in a light guide to the photomultiplier. The photon is again converted into a current of electrons, and a positive potential can attract the electrons and collect them, and they will be detected. Figure 4.5 shows operator focus conditions on the sample: short working distance (left) and long working distance (right).

4.4 Microscopy Techniques

Fig. 4.5 Test focus radius path: short working distance (left) and long working distance (right)

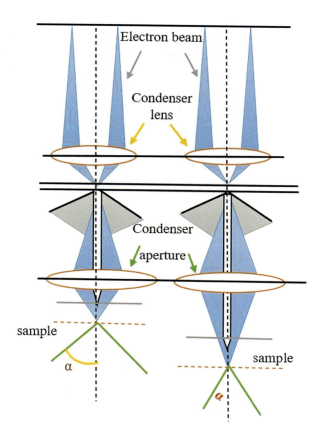

When the sample is moved at a greater distance from the lens, several phenomena occur like, working distance increases; the demagnification decreases; the spot size increases; the divergence angle α decreases.

By decreasing demagnification, the focal length of the lens will increase.

The resolution of the specimen decreases with increasing working distance, and the spot size increases. Accordingly, the depth of field increases with increasing working distance because the divergence angle is smaller [15–17]. Figure 4.6 presents different contrast mechanisms for SEM images secondary electrons and scattered electrons for sample ZnSe.

4.4.2.1 Energy Dispersive X-Ray Spectroscopy (EDS)

The capabilities of chemical characterization of the sample by this method are due to the fundamental principle that each element of the periodic table is a unique electronic structure and thus a unique response. When a beam of electrons or photons is directed towards a sample, it produces a measurable response that characterizes the

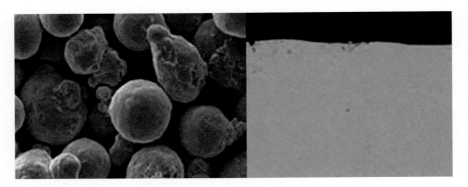

Fig. 4.6 Backscattered Electron SEM Imaging for ZnSe sample

specimen. Electromagnetic waves stimulate this response. When at rest, the electrons in an atom within the sample are unexcited and situated primarily in layers concentrated around the nucleus.

The incident beam excites an electron from one of the inner shells of the atom, causing it to eject and contributing to the formation of an electronic void inside the electronic structures of the atom. When an electron from a higher-energy shell fills the void in a lower-energy shell, any excess energy is released in the form of X-rays.

X-ray emission creates spectral lines that are specific to individual elements. Thus, a spectrum can be obtained by this technique in which peaks correspond to specific X-ray lines. Quantitative data can also be obtained by comparing the height or area of the peaks in the X-ray spectrum of an unknown material with those of a standard material.

Excess energy from an electron migrating to an inner shell (to fill our created void) can do more than emit X-rays. Often instead of emitting X-rays, the excess energy is transferred to a third electron from a more distant outer shell, causing its ejection. This species of the ejected electron is called the Auger electron, and the method for its analysis is known as Auger Electron Spectroscopy (AES). Information on the amount and kinetic energy of ejected electrons is used to determine the binding energy of these freshly released electrons, which is specific to each element and thus the sample's possible chemical characterization. The spectrometer transforms the energy of every X-ray into a voltage signal proportionate in magnitude. This phenomenon is achieved following a three-stage process.

First, the X-ray is transformed into an electric charge by ionization in the semiconductor crystal. After that, this load is converted into a voltage signal by the FET preamplifier.

Finally, the voltage signal is input to a pulse processor to measure the signals and then passed to an analyzer that displays and analyzes the data. The signal from the preamplifier output is a voltage "ramp," where each X-ray appears as a voltage step within that ramp Fig. 4.7.

Foinard et al. [18] studied in vitro assessment of the interaction between amino acids and copper in neonatal parenteral nutrition. The scanning electron microscopy

4.4 Microscopy Techniques

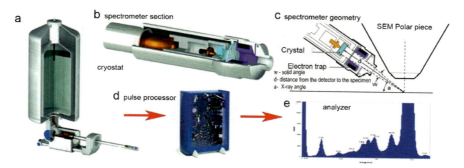

Fig. 4.7 Detailed presentation of an X-ray spectrometer (**a**) Cryostat (**b**) Spectrometer Section (**c**) Spectrometer Geometry (**d**) Pulse Processor (**e**) Analyzer

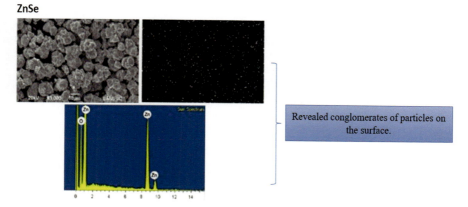

Fig. 4.8 SEM-EDS images of ZnSe layer

(SEM) investigations were carried out on a Hitachi S-4700 SEM FEG (Field Emission Gun) (Hitachi, Vélizy-Villacoublay, France) equipped with an energy dispersive spectroscopy (EDS) microanalysis system. Individual particles were selected for elemental analysis using a Noran System Detector with an ultra-thin window (Thermo Electron Corporation, Waltham, MA). Figure 4.8 shows the analyzed ZnSe layer and revealed conglomerates of particles on the surface. EDS analysis identifies various elements of a conglomerate. This is a qualitative method of investigation.

4.4.2.2 Electron Backscatter Diffraction (EBSD)

Electron backscatter diffraction (EBSD) is also known as Backscatter Kikuchi diffraction (BKD), and it is a microstructural crystallographic technique used to study the crystal structure or preferential orientation of any crystalline or polycrystalline material. EBSD is used to index and identify seven crystal systems and is applied to mapping crystal orientation, defect studies, phase identifications, morphology studies, investments in regional heterogeneity, material identification, and using complementary techniques, physical-chemical identification. EBSD analysis is performed experimentally using an SEM equipped with a scattering electron recording camera. In essence, the diffraction chamber is a CCD focal plane device coupled to a phosphor screen inserted into the SEM test chamber at an angle of 90° compared to the polar piece. A charge-coupled device (CCD) is an image sensor consisting of an integrated circuit containing a network of photosensitive capacitors interconnected. A flat/polished crystalline specimen is placed in the test chamber at a large inclination angle (70° from the horizontal direction) to the diffraction chamber, satisfying Bragg's conditions, turning into scattered diffraction. Due to the inclined angle of placement to the specimen, these diffracted electrons are directed to the phosphor screen at the diffraction chambers they hit, causing its fluorescence; the CCD then detects this fluorescent light. The diffracted electrons form an image completed by a diffraction spectrum in the diffraction chamber, unique for the microstructural-crystallographic properties of the material. Each diffraction pattern will have several intersecting lines called Kikuchi bands, which correspond to each of the diffraction planes of the networks and can be indexed individually using the Miller indicators (*hkl*) of the diffraction planes. Zecevic [19] studied the residual ductility and microstructural evolution in the continuous-bending-under-tension of AA-6022-T4. The EBSD data collection was performed using the Pegasus system (Octane Plus SDD detector and Hikari High-Speed Camera) attached to a Tescan Lyra (Ga) field emission SEM at a voltage of 20 kV. The EBSD scans were run with either 1.5 or 2 µm step size. Figure 4.9 shows the SEM image of the microstructure and texture of the sample ZnSe.

4.4.3 Transmission Electron Microscopy (TEM)

Electron diffraction is one of the most used methods to obtain information on solid, crystalline, or amorphous materials. This method complements other methods used in material identification, such as X-ray diffraction, atomic spectroscopy, and FT-IR.

Quantitative information that can be obtained by electronic diffraction includes:

- Determining the interplanar distances in the crystal;
- Determining the crystallization in the material;
- Identification of the phases and the relationship between the orientation of these towards the material matrix;
- The crystallographic description of the defects produced by deformations, irradiations.

Although optical microscopy has been known for over 400 years, electron microscopy is a relatively recent technique, first appearing in 1931. Over the years, the resolution of electron microscopes has developed substantially from 5 nm (in 1949 the first commercial electron microscope was Philips EM 100) [20]

4.4 Microscopy Techniques

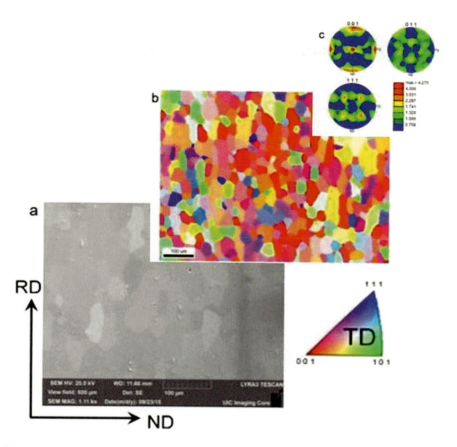

Fig. 4.9 Microstructure (**a**) and texture (**b**), (**c**) for ZnSe sample

to low electronic levels (Philips CM 300-Ultra TWIN equipped with cannon with field emission field resolution 0.17 nm and information limit 0.10 nm) [21].

Long before the electron microscope was invented, Ernest Abbé, a German physicist, established the first theory for the optical microscope, taking into account the undulating nature of light. Although microscopists believed that the quality of the lenses limited the resolution of the optical microscope, Abbé established in 1976 based on his theory that the nature of light (the long wavelength) was the limiting factor. At that time, he already believed that "… processes still unknown… can cross borders" without even knowing of electronic existence.

To see how microscopy has evolved, we can mention four of the most important events that led to today's interdisciplinary science: electrical microscopy. In 1879 Abbé completed the theory for optical microscopy and discovered that wavelength is a factor that limits the resolution in optical microscopy. In 1897 J.J. Thomas announced the existence of negatively charged particles, later called electrons. In

Fig. 4.10 Operating diagram for a TEM electron microscope

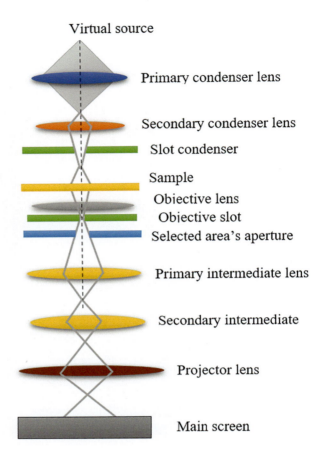

1931 the first transmission electron microscope was built by a team led by Ernst Ruska, and in 1949 Philips launched the first commercial version of an electron microscope, EM100, with a 5 nm resolution, electromagnetic lenses, and a special device called a "wobbler" [22].

General Operation Description of a Transmission Electron Microscope
The functional diagram of the electron microscope is presented in Fig. 4.10.

The *virtual source* is represented by an electronic cannon that produces a monochromatic flow of electrons.

This flow is focused coherently in the thin beam by using *primary* and *secondary condenser lenses*. The primary lenses determine the size of the spot.

The secondary lenses change the size of the spot when it encounters the sample, changing it from a vast flux dispersed in a point beam. The capacitor slot limits the flow. The electronic flow is transmitted on the sample and is focused by the objective lens into an image.

4.4 Microscopy Techniques

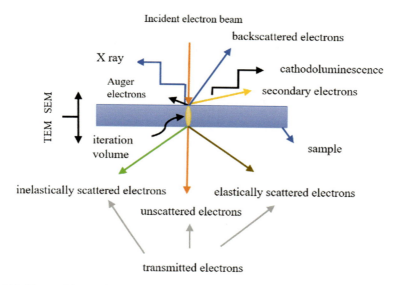

Fig. 4.11 Types of interactions in the sample analyzed with the electron microscope

The objective slit increases the contrast by blocking the diffracted electrons at a large angle. The slit (aperture) of the selected area allows the user to examine the periodic diffraction of electrons through the ordered structure of the atoms in the sample.

The image is moved down on the column through the *intermediate and projector lenses*, thus enlarging after each cycle. The image hits the main screen, generating light, allowing the user to see the image. The darker areas of the images show those regions in the sample through which fewer electrons were transmitted (the electrons are more compact, denser) [22, 23]. The brighter areas of the images are regions of the sample through which more electrons were transmitted (they are less dense).

Electronic Interactions in the Sample Analyzed
When the sample is "hit" by the flow of electrons, many reactions occur inside it. The reactions are shown at the top in Fig. 4.11 for SEM analyses, and at the bottom are used in the TEM analyses.

Interactions in Large Thickness Samples
- *The scattered electrons* appear due to the collision of the incident electrons with an atom in the sample, the angle between the scattering electron path and the incident electron path being very small (so the incident electron is diffused back by almost 180°).

The creation of scattered electrons varies directly with the atomic number in the samples. This different production rate causes elements with a higher Z to appear brighter than those with a lower atomic number. This interaction is used to differentiate the sample components with different Z. The detection of scattered electrons is done using a solid semiconductor device mounted in the lower end of the objective lens. When a scattering electron hits this detector, electron-empty pairs are created, which are then counted. This amount is transformed into a certain intensity of pixels and presented on the CRT screen (cathode ray tube), forming the image [23].

- *Secondary Electrons*

An e^- incident produces a secondary electron following its very close passage past an atom in the sample, close enough to divide some of its energy into a lower energy electron. This causes a slight loss of energy and a change of path of the incident electron, and the ionization of the electron in the sample atom. This ionized electron then leaves the atom with a very low kinetic energy (5 eV) called the *secondary electron*. The production of the secondary electron is closely related to topography. Due to their low energies, only secondary electrons that are very close to the surface (at 10 nm distances) can leave the sample and be examined. Any change in the topography of the sample more significant than a depth of 10 nm will change the secondary electrons' yield due to the collection efficiency. The collection of these electrons is done by using a collector directly connected to the detector for secondary electrons.

This collector is a network subject to a potential of about 500 V and is placed in front of the detector, attracting negatively charged secondary electrons, which will pass through the network holes in the detector to be counted.

- *Auger Electrons*

Because lower electron energy was emitted from the atom during the emergence of the secondary electron, an internal low-energy level will now have a vacancy. A higher energy electron from the same atom can "fall" to a lower energy level, occupying that vacation. This phenomenon causes the appearance of an energy surplus in the atom that can be corrected by emitting an external, low-energy electron called an *Auger electron* [24, 25].

Auger's electrons have characteristic energy unique to each element from which it was issued. These electrons are collected and sorted by energy to give compositional information about the sample.

Interactions in Small Thickness Samples

Scattered electrons are those electrons transmitted directly through the thin specimen without any interaction occurring inside it. The transmission of the scattered electrons is inversely proportional to the thickness of the sample. The thicker areas of the sample will have less scattered electron transmission and thus will appear darker. The thinner areas will have a higher concentration of scattered electrons and thus appear brighter [25].

The elastically scattered electrons are incident electrons scattered by the atoms in the sample elastically (without loss of energy). These electrons are transmitted through the unexplored areas of the material. Electrons comply with Bragg's law and are scattered with a wavelength equal to the product of twice the distance between atoms and the sine of the scattering angle. Incident electrons have the

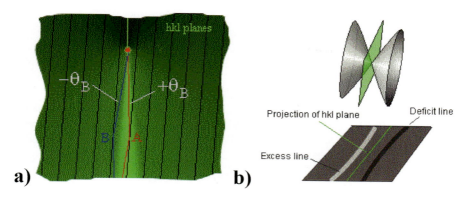

Fig. 4.12 Characterization of a Kikuchi line: (**a**) construction of the Kikuchi line, (**b**) Cones Kikuchi [26]

same energy (and wavelength) and enter the specimen in a perpendicular direction to the surface, and therefore will be scattered at the same angle. These scattered electrons can be collimated using magnetic lenses to form a pattern of spots (each spot corresponding to given certain interatomic distances). Therefore, this model can provide information on orientation, atomic arrangement, and phases present in the area under examination. Magnetic lenses are circular electromagnets capable of projecting a precise circular magnetic field in a specific region. The field behaves like optical lenses, having the same characteristics (focal length, divergence angle) and errors (spherical, chromatic aberration).

The inelastically scattered electrons are electrons that interact inelastically with sample atoms losing energy during the interaction [26].

There are two ways to use these electrons:

1. *Electron Energy Loss Spectroscopy* – the inelastic energy loss of incident electrons is characteristic of the elements they interact. These energies are unique to each bond state of each element. Thus, they can be used to extract compositional information and the bond between the elements (e.g., the oxidation state).
2. Kikuchi bands are bands of alternating light and dark lines formed by inelastic scattering interactions concerning the interatomic spaces in the sample. These bands can be measured (their width is inversely proportional to the interatomic distance) or followed by comparative study (in the form of a map) with models of elastically scattered electrons.

Kikuchi Lines

Electrons that have been inelastically scattered can be diffracted later, but only if they move at the Bragg θ_B angle associated with a set of planes. Two sets of electrons will be able to do this: those at $+\theta_B$ and those at $-\theta_B$ (see Fig. 4.12b). This diffraction is manifested by changes in light intensity in the background.

Because there are more electrons in position A than in position B (electrons moving through position A are closer to the incident direction than those in B), a single light line (excess line) is developed next to a dark line (deficit line) (see Fig. 4.12b). Because electrons are inelastically scattered in all directions, diffracted electrons will form a cone, not a beam.

This way, we will be able to observe Kikuchi lines and not Kikuchi spots.

The space occupied by a pair of Kikuchi lines is the same as diffracted spots in the same plane. The orientation of the specimen very sensitively controls the position of the lines, and Kikuchi lines are often used to determine the orientation of a crystal in TEM analysis, with an accuracy of 0.01° [24–26].

Transmission Electron Microscopy (TEM) Applications

Transmission electron microscopy is widely used in the field of materials study. The samples that need to be analyzed need to be resistant to the vacuum advanced from the microscope. They must be prepared in the form of thin sheets to be penetrated by the electron beam. Preparation techniques for obtaining electronic transparent regions include ion grinding of the sample by ion beam and angle grinding. The focused ion beam (FIB) technique is relatively new for preparing thin samples for TEM examination, starting from larger thickness samples. Materials that are small enough to be transparently electronic, such as powders or nanotubes, can be quickly prepared for analysis by depositing a dilute sample containing the specimen on support networks. The suspension is usually a volatile solvent, such as ethanol, allowing the solvent to evaporate rapidly so that the sample can be readily analyzed. Defects in crystals can affect both the properties of the mechanism and the electronic properties of the materials. So understanding their behavior provides a better understanding of the property of the sample. If the sample is oriented when a particular plane is only slightly inclined with respect to the strongest diffraction angle (Bragg angle), any distortion in the crystal that locally inclines that plane to the Bragg angle will produce extreme contrast variations of the image. Defects that produce only displacements of atoms that do not tilt the crystal plane to the Bragg angle (e.g., displacements parallel to the crystal planes) will not produce a strong contrast.

The use of electrons offers various advantages:

- Better resolution due to the short wavelength compared to light;
- The low mass of electrons leads to nondestructive interactions in the examined sample;
- Simple focusing of the beam by changing the current in the lens, and following the study of the interaction electron sample, a series of useful signals appear such as direct TEM image;
- Qualitative examination and spectrographic analysis of the sample by capturing the X-rays emitted after the electron bombardment of the specimen and not finally the electron diffraction.

All these advantages are accompanied by the need to maintain a high vacuum in the column and to examine samples of sufficiently small thickness to be traversed by the electron beam.

4.4.3.1 High-Resolution Transmission Electron Microscopy (HRTEM)

High-resolution transmission electron microscopy (HRTEM) allows direct observation of crystalline structures. One advantage of this method of investigation is that there is no correlation between the contrast variations in the image and the location of the defect. However, it is not always possible to directly interpret network images in terms of structural compositions of the sample because the image is sensitive to a number of factors (sample thickness and orientation, defocus of objective lenses, spherical and chromatic aberration); although a quantitative interpretation of the contrast appearing in network images is possible, it is naturally complicated and necessary or extensive simulation of images. Computer simulation of these images has brought a new level of understanding in the study of crystalline materials.

In HRTEM analysis, the image formation is based on phase contrast: the atoms in material diffract the electrons as they penetrate the material (the relative phases of the electrons change as a result of their transmission through the sample), causing a diffraction contrast that adds to contrast already present in the transmission electron beam. Through the complex combination of the contrasts to the multiple diffraction planes and the transmitted electron beam, the computer performs simulations used to determine the type of contrast produced by various structures. Thus, it is necessary to know a high level of information about the material that needs to be analyzed before a "phase-contrast" image can be interpreted correctly.

Phase-contrast images can be formed by completely removing the objective lens, or using a very large objective lens. By this action, we ensure that not only the transmitting beam but also the diffracted electrons contribute to the formulation of the image. In order to form the image in this analysis, the contrast that appears from the interference of the electron wave with itself in the plane image is used. Due to the inability to record the phases of these waves, the amplitude resulting from this interference is generally measured. This is only valid if the sample is thin enough so that the amplitude variations do not contribute to the image.

4.4.3.2 TEM Automated Crystallography – ACT

Peak detection is a key step in the automatic indexing of diffraction patterns. ACT provides an algorithm for finding peaks and measuring their position in the diffraction space relative to the model's center. To understand how this algorithm works, we will first consider a diffraction image as a set of values (intensities) located in particular coordinates $(X–Y)$. This can be seen in the Fig. 4.13. Using a mobile "mask" centered on each pixel, the image is divided into small areas which allow a pixel to be compared only to pixels within the same small area. The next step is to draw a Gaussian curve for pixel intensities and calculate the "volumetric" center of the peak.

The user can eliminate noise and lower peaks from an algorithm by setting the peak's height and inclination and the mobile mask's size.

Fig. 4.13 ACT algorithm for a set of values (intensities) located in particular coordinates (X–Y)

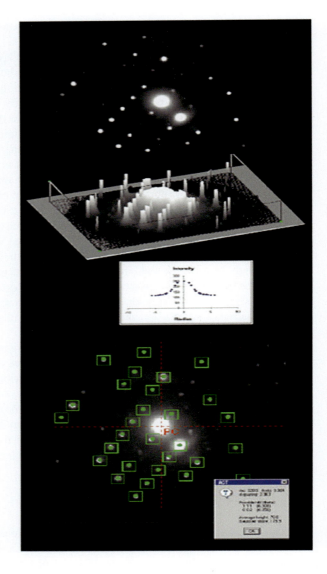

The maxima are indicated by green squares and the centers of the peaks are indicated by the point inside the squares. Each spot can be analyzed manually to find specific details such as position, height, and possible (*hkl*) planes [22].

TSL [23] has developed an automatic indexing procedure to determine whether the orientation best fits a given diffraction spot pattern. The various steps involved in the indexing procedure are presented below. For each spot detected in the model, a list of the potential corresponding plane (*hkl*) is determined by comparing the projection of the length of the diffraction vector in relation to a tabulated list (see

4.4 Microscopy Techniques

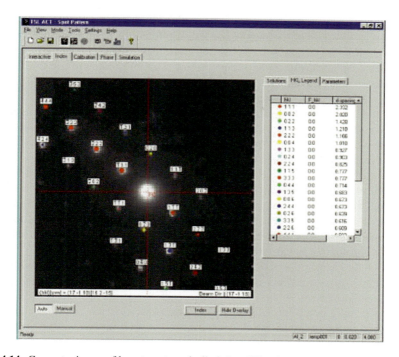

Fig. 4.14 Computer image of how to automatically index diffraction spots [26]

Fig. 4.14). To reduce the list of potential planes (*hkl*), the angles between a triplet of spots are compared with a table of interplanar angles. The remaining planes (*hkl*) are used to have a general set of spot triplet orientation solutions, which supposes that the spots are on an Ewald sphere.

Similar orientations are grouped and can be calculated or averaged. Average records appear for variations in point positions in the model. These variations occur when the points in the reciprocal networks are slightly displaced from the exact intersection with the Ewald spheres. This technique provides automatic indexing for spot models that are both inside and outside the axes. The user can control the critical adjustment and tolerance parameters to optimize the indexing process to a given material. This procedure is suitable for overcoming many difficulties encountered during the automatic indexing of the models of their diffraction spots [24–26].

Both amplitude phase and contrast participate in image formation in HR-TEM mode. At small magnifications, below 100,000×, the amplitude contrast predominates. At very large magnifications, the phase-contrast approaches or even exceeds that in amplitude and participates in the formation of the image. The images obtained on crystalline samples are generally formed by recombining the primary beam with one or two diffracted planar beams. Due to diffraction, there is a strong phase variation in the image plane. Periodic fringes are thus formed due to the interference phenomenon. Under certain conditions, these fringes correspond to the crystalline plane participating in the diffraction. In general, the visibility of the fringes depends only on the thickness of the sample and the fulfillment of Bragg's conditions. Also, the fringes do not coincide with the atomic planes if they form images on a marginal area where the thickness of the sample varies. This phenomenon can be identified relatively easily due to the curvature of the fringes. Another factor that influences the formation of fringes is astigmatism given by the objective lens.

HR-TEM images can penetrate the crystal structure. The lines in the image are a direct consequence of the electronic diffraction on the crystalline planes whose orientation verifies Bragg's law. There are situations very useful in the complete characterization of the crystal, in which the orientation of the crystal allows diffraction on several planes. In the same case, the exact area axis can be established, i.e., it had been obtained by the intersection of the plane. Thus, with such an important landmark, the calculations regarding the determination of the crystalline structures of the sample can be carried out. Determining crystalline structures is one of the most challenging and delicate operations in crystallography. Thus, many phases belong to the same substances, but we have certain different parameters.

The images inserted in HRTEM are the complementary images in the Fourier space. These images are very useful for accurately measuring the distances between lines. The reason is that in the Fourier space, each line in the real image corresponds to a point, more precisely a set of points symmetrical to the origin of the image established in the center of these. The distance between the lines is determined by measuring the distance between the information points. The accuracy of the measurements increases because locating a point is much easier than locating a point in the direct image with a line [25, 26].

In this high-resolution image, one of the nanocrystals is thus oriented and allows the occurrence of the interference phenomenon. Thus, visible fringes appear that come from the diffraction of the electron beam on the plane. Figure 4.15 is the Fourier representation of the image that is used to measure distances. The arrow indicates fringe applicator points.

4.4.3.3 Selected Area Electrons Diffraction – SAED

A second operation mode of an electron microscope by transmission is electron diffraction. When the sample is crystalline, several diffracted beams with small 2θ angles occur, all of which converge in the focal plane of the object's images. A diffraction pattern is formed in this plane corresponding to a flat section of the networks, reciprocals of the crystal. A more intense continuous background is superimposed in the center, coming from the inelastic diffusion. There are two types of diffraction:

- Microdiffraction: the area of participation in diffraction with ($\Phi = 1$ μm) is selected at the level of the object image plane using a "field selector" diagram.
- Nano-diffraction: the direct active area is selected using an excellent incident beam ($\Phi = 1$ nm).

4.4 Microscopy Techniques

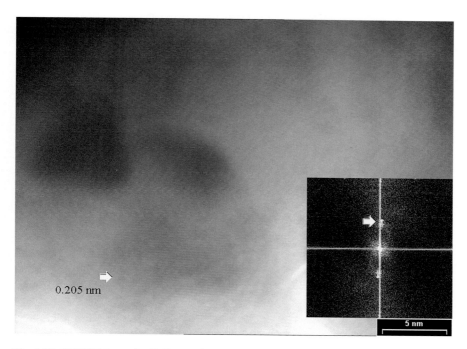

Fig. 4.15 HRTEM image for ZnSe sample

Electron diffraction is an elastic scattering phenomenon, with electrons scattered by atoms in a regular network (crystal).

The plane wave of the electron interacts with the atoms and generates secondary waves that interfere with each other (analogous to the Huygens principle for light diffraction). This occurs constructively (hardening at safe scattering angles generates diffraction rays) or destructive (extinguishing the beam).

Similar to X-ray diffraction, scattering can be described as a reflection of rays on the planes of atoms (the planar lattice). With the help of Bragg's laws from the templates, one can calculate the interplanar distances by knowing the wavelength of the electron. More information on crystal symmetry can be obtained. Consequently, electron diffraction is a valuable tool in crystallography.

Electron diffraction is especially well suited to polymers, which have imperfect crystallization and do not form macroscopic crystals.

The field of application of this technique is vast. It extends to the measurement of crystallographic parameters, determination of crystalline structures, characterization of polymorphisms and phase transitions, study structures, the orientation of single crystals, and thin films. D. Wei et al. [27] studied the morphology and structure of the BiOCl and Ag/AgCl/BiOCl ternary composites characterized by SEM and TEM images. The images gave information about crystallography and various groups of diffraction spots or rings, indicating the production of composites. Figure 4.16 shows the SEM and TEM images for ZnSe sample.

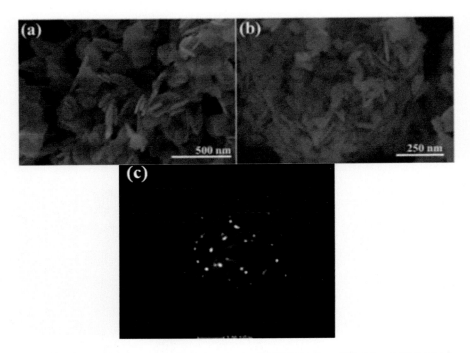

Fig. 4.16 SEM and TEM images for ZnSe sample; (**a**) SEM images of S1 (**a**) and S3 (**b**), high-resolution TEM (**c**) of S3 products

References

1. Todd C. Hufnage, *Structural characterization of amorphous materials using x-ray scattering*, ebook, pg. 7–14, (2002).
2. C. K. Saw, T. Lian, S. D. Day, J. C. Farmer, *X-ray Diffraction Techniques for Structural Determination of Amorphous Materials*, ebook, pg. 3–10, (2006).
3. I. G. Murgulescu, *Introducere în chimia fizică, pg. 10–35*, vol. I, Ed. Academiei RSR, București, (1976).
4. I. G. Murgulescu, *Introducere în chimia fizică, pg. 25–37*, vol.,2, Ed. Academiei RSR, București, (1979).
5. M. Fabian, N. Dulgheru, K. Antonova, A. Szekeres, M. Gartner, *Investigation of the Atomic Structure of Ge-Sb-Se Chalcogenide Glasses*, Advances in Condensed Matter Physics (2018).
6. http://www.mdeo.eu/MDEO/Studenti/Docs/AFM_seminar_2011.pdf
7. M. Bozgan, *Istoria Ochelarilor*, p. 62, Revista, Istorie și Civilizație, Nr. 2, Nov, (2009).
8. Y. Oshikane, T. Kataoka, M. Okuda, S. Hara, H. Inoue & Motohiro Nakano, *Observation of nanostructure by scanning nearfield optical microscope with small sphere probe*, Sci. Technol. Adv. Mater. 8: 181 (2007).
9. Nonnenmacher M., M. P. O'Boyle and H. K. Wickramasinghe, *Kelvin probe force microscopy*, Appl. Phys. Lett. 58, 2921, (1991).
10. Y. Uehara, J. Michimata, S. Watanabe, S. Katano, and T. Inaoka, *Determining the phonon energy of highly oriented pyrolytic graphite by scanning tunneling microscope light emission spectroscopy*, Journal of Applied Physics 123, 104306 (2018).

References

11. Dulgheru (Nedelcu) N, *Correlation of optical and morph-structural properties in chalcogenide compounds with applications in optoelectronics*, PhD thesis, Romanian Academy, 2019.
12. S. Baraitareanu, *De la microscopul optic la microscopul de forta atomica*, pg. 444–452, Editura Tehnica (2011).
13. G. Binnig, H. Rohrer, *Scanning Tunneling Microscopy from Birth to Adolescence*, pg. 606–614, Angewandte International Ed. Chemie (1987).
14. Bojin D, Vasiliu F, *Microscopia electronica*, pg 75–86, Editura Științifică și Enciclopedică București (1986).
15. Chicinaş I., A. Molinari, *Noi tehnologii pentru materiale avansate*, pg. 23-117, Ed. Press, (1997).
16. M. Nonnenmacher, M. P. O'Boyle and H. K. Wickramasinghe, *Kelvin probe force microscopy*, Appl. Phys. Lett. 58, 2921, (1991).
17. R. Jafari, E. Sadeghimeresh, T. S. Farahani, M. Huhtakangas, N. Markocsan, S. Joshi, *KCl-Induced High-Temperature Corrosion Behavior of HVAF-Sprayed Ni-Based Coatings in Ambient Air*, Journal of Thermal Spray Technology 27(3):500–511.
18. A. Foinard, M. Perez, C. Barthélémy, D. Lannoy, F. Flamein, L. Storme, A. Addad, M.A. Bout, B. Décaudin, P. Odou, *In Vitro Assessment of Interaction Between Amino Acids and Copper in Neonatal Parenteral Nutrition*, Journal of Parenteral and Enteral Nutrition 40(6), February 2015.
19. M. Zecevic, T. J. Roemer, M. Knezevic, Y. P. Korkolis B. L. Kinsey, *Residual Ductility and Microstructural Evolution in Continuous-Bending-under-Tension of AA-6022-T4*, Materials 9(3):130, February 2016.
20. Rudenberg, Reinhold (May 30, 1931), *Configuration for the enlarged imaging of objects by electron beams*, Patent DE906737.
21. R. Wang, J. Tao, K. Du, Y. Wang, B. Ge, F. Li et al. *Progress in Nanoscale Characterization and Manipulation*, Pages 69–203, Springer 2018.
22. F. Krumeich, *Introduction into transmission and scanning Transmission Electron Microscopy*, pg. 3–5, 8093 Zürich, Switzerland.
23. F. E. Rauch, J. Portillo, S. Nicolopoulos, D. Bultreys, S. Rouvimov, P. Moeck, *Automated nanocrystal orientation and phase mapping in the transmission electron microscope on the basis of precession electron diffraction*, Z. Kristallogr. 225 (2010) 103–109.
24. S. Zaefferer, *Computer-aided crystallographic analysis in the TEM*, Adv. in Imaging and Electron Physics, Vol. 125, 2003, Pgs. 355–415, II–XII.
25. J. Jeong, N. Cautaerts, G. Dehm, C. H. Liebscher, *Automated Crystal Orientation Mapping by Precession Electron Diffraction-Assisted Four-Dimensional Scanning Transmission Electron Microscopy Using a Scintillator-Based CMOS Detector*, Microscopy and Microanalysis (2021), 27, 1102–1112.
26. N. Dulgheru, *CuNiFe thin layers, obtaining and characterization*, Dissertation thesis, Univ. Ovidius Constanta, 2009.
27. D. Wei, F. Tian, Z. Lu, H. Yang, R. Chen, *Facile synthesis of Ag/AgCl/BiOCl ternary nanocomposites for photocatalytic inactivation of S. aureus under visible light*, May 2016, RSC Advances 6(57).

Chapter 5
Optical Analysis and Chemical Properties

Abstract The chemical structure was analyzed by spectrophotometer measurements and models were developed based on the Swanepoel method using transmittance, absorbance, and reflectance spectra. The Swanepoel model obtained the optical constants (n – refractive index, and k – extinction coefficient) and the thickness (d).

Keywords Optical and chemical properties · UV-VIS spectroscopy · Spectral transmission · Reflections spectra · Swanepoel method · Refractive index · Extinction coefficient · Band-gap · Urbach energy · Absorption coefficient · Dielectric constant · OpenFilters · Antireflection coating · Optical conductivity · Spectro-ellipsometry method · Infrared spectroscopy (FT-IR) · Raman spectroscopy

The chemical structure was analyzed by spectrophotometer measurements and models were developed based on the Swanepoel method using transmittance, absorbance, and reflectance spectra. The Swanepoel model obtained the optical constants (n – refractive index, and k – extinction coefficient) and the thickness (d).

UV-VIS-NIR and IR spectrum ellipsometry (SE) – it is a nondestructive indirect optical technique that allows the characterization of thin films, surfaces, and interfaces used to determine the thickness of thin films, optical constants, band-gap, and electrical parameters (mobility, conductivity, carrier density).

Ellipsometry does not directly measure the film thickness or optical constants. The analysis of ellipsometry is performed using an optical model. This model is an approximate structure of the samples and includes the order of layers for this material, optical constants, and the thickness of the layers. When these are not known, it starts from a probable value. The fitting stage is represented by using a dedicated calculation program for generating data.

The most used fitting parameters are thickness and optical constants. In some cases, the optical constants can be found listed in the literature. The experimental data can compare the result obtained by spectrophotometry and spectro-ellipsometry measurements.

© The Author(s), under exclusive license to Springer Nature Switzerland AG 2023
N. Nedelcu, *Thin Films*, https://doi.org/10.1007/978-3-031-06616-0_5

5.1 Study of Optical and Chemical Properties

5.1.1 UV-VIS Spectroscopy

Spectroscopy in the visible and ultraviolet field is one of the modern physical methods for studying the structure and properties of thin films. The associated electronic transitions in the visible and ultraviolet spectra affect moving the molecule from the ground state to an excited state. The spectra are called electronic spectra and they give information on the electronic state of the molecules [1]. An electronic spectrum is the energy absorption curve. In atomic or molecular systems, electronic spectra result from quantum transitions resulting from the absorption or emission of light energy from the visible or ultraviolet range. Thus, the transition of the respective system from a lower energy level to a higher energy level, respectively, in an excited state, will take place only if an atomic or molecular system absorbs energy from a particular spectral range. The atom or molecule remains excited only approx. 10.8 s, after which it returns to the initial state (fundamental state), emitting radiation of the same energy as the absorbed one. This radiation gives a line in the absorption spectrum of the analyzed system.

This process of forming spectral lines can be analyzed using a spectrophotometer device. The spectra representation is in the absorption A, transmission T, or R reflection, depending on ν (wavenumber) or λ (wavelength) [1, 2].

Nedelcu [3, 4] developed models based on the Swanepoel method using transmittance, absorbance, and reflectance spectra (Fig. 5.1). The spectra were measured using Lambda 950 Spectrophotometer. According to the Swanepoel model [5] (described in Sect. 5.2.2), the optical constants (n – refractive index, and k – extinction coefficient) and the thickness (d) were obtained (Fig. 5.2). The experimental data recorded by a spectrophotometer can calculate the absorption coefficient, band-gap, Urbach energy, optical conductivity, and dielectric constant. The optical properties are the most crucial characteristics in defining materials. In optical coating production departments, the optical constants and thickness are primarily determined, to create a new material.

Many engineers use various software created especially to design a material with technological applications in different industrial domains. Software like OptiLayer, TFcalc, Strat, and OpenFilters can be used for optical coating design. Having the optical properties can add the material in the bank of materials and design a new material based on thickness simulated [7]. The new material is created depending on the optimization condition. The optimization condition is represented by high/low transmission or high/low reflection.

Nedelcu [8] uses OpenFilters software to create a highly transparent conductive optical coating optimized for an oblique angle of incidence based on optical properties determined by the Swanepoel method recorded on a Spectrophotometer. The antireflection coating consists of six thin layers (three layers of MgF_2 and three layers Ti_3O_5). Fig. 5.3a displays the theoretical spectral reflection factor, calculated by OpenFilters, optimized for an angle of incidence between 0 and 30° and for a

5.1 Study of Optical and Chemical Properties

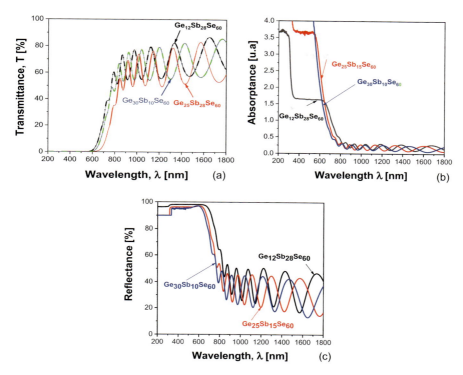

Fig. 5.1 Spectral measurements by Lambda Spectrophotometer, (**a**) spectral transmission [3, 4], (**b**) spectral absorptance [4], (**c**) reflectance spectra [4]

spectral range of 450–950 nm. Fig. 5.3b, on the other hand, shows the experimental results obtained by Spectrophotometer Lambda 950.

The optical absorption coefficient α can be calculated in terms of the transmittance spectra T and reflectance spectra R by using the following relation:

$$\alpha = \frac{1}{d} \ln \left[\frac{(1-R)^2}{T} \right] \quad (5.1)$$

where α is absorption coefficient, R and T are reflectance and transmission spectra. If we do not have T and R spectra and if we have absorbance spectra A, then the absorption coefficient can be calculated:

$$\alpha = 2.303 \frac{A}{d} \quad (5.2)$$

where A – absorbance and d – thickness of the thin film. The relationship between the absorption coefficient and the incident photon energy $h\nu$ is governed by the relation [9, 10]:

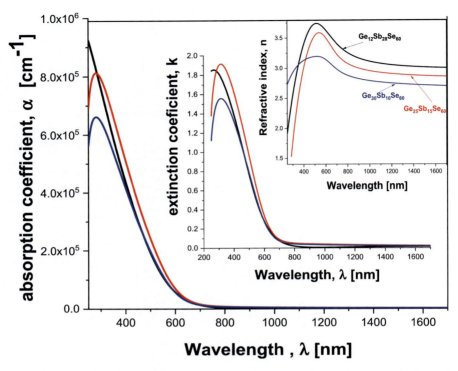

Fig. 5.2 The optical constant determined by Swanepoel method: refractive index, n, extinction and absorption coefficient (k) [3, 6]

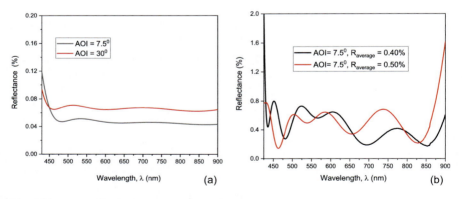

Fig. 5.3 The antireflection coating optimized for an angle of incidence AOI = 0–30° calculated by OpenFilters (**a**) and obtained experimentally by Spectrophotometer Lambda 950 (**b**)

$$\alpha h\nu = A\left(h\nu - E_g\right)^n \qquad (5.3)$$

while ($h\nu$) is the proton energy:

5.1 Study of Optical and Chemical Properties

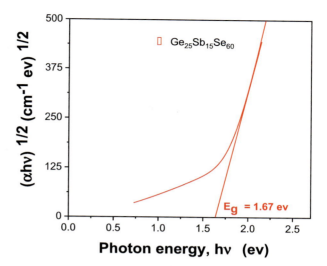

Fig. 5.4 The E_g calculated by Tauc formula exemplified for $Ge_{25}Sb_{25}Se_{80}$

$$h\nu(\text{eV}) = \frac{1240}{\lambda(\text{nm})} \qquad (5.4)$$

If we plot a graph between $(\alpha h\nu)^{1/n}$ and $(h\nu)$, Fig. 5.4, we can get a straight line. The values of E_g, estimated from the intercept where $(\alpha h\nu)^{1/n} = 0$, are given by the point where this line intersects the X-axis. The value of n is dependent on the electronic transition type, where $n = \frac{1}{2}$ for direct allowed transition, $n = 2$ for indirect allowed transition, $n = 3$ for direct forbidden transition, and $n = \frac{3}{2}$ indirect forbidden transition. It is necessary to select the suitable n according to your samples and their preparations. The value $Ge_{25}Sb_{15}Se_{60}$ was determined using the Tauc formula applied to the experimental data obtained from the spectrophotometer measurements in Fig. 5.1, as illustrated in Fig. 5.4.

The optical band-gap E_g was derived assuming indirect transitions between the edge of the valence and conduction band. Nedelcu et al. [4] used six models to determine the optical band-gap energy. These models were based on the absorbance and absorption coefficient. The optical energy was determined by extrapolation of the linear curve of A, $A^{\frac{1}{2}}$, α, $\alpha^{\frac{1}{2}}$, $Ah\nu^{\frac{1}{2}}$, and $\alpha h\nu^{\frac{1}{2}}$, as a function of the photon energy $h\nu$. The intersection of the linear region on the $h\nu$ axis gives E_g, Fig. 5.5.

There is an exponential part called the *Urbach tail* along the absorption coefficient curve and near the optical band edge. The spectral dependence of the absorption coefficient (α) and photon energy $h\nu$ is known as Urbach empirical rule [11], which is given by the following equation

$$A = \widetilde{A} \cdot e^{\left(\frac{h\nu}{E_u}\right)}$$

$$\alpha = \widetilde{\alpha} \cdot e^{\left(\frac{h\nu}{E_u}\right)}, \qquad (5.5)$$

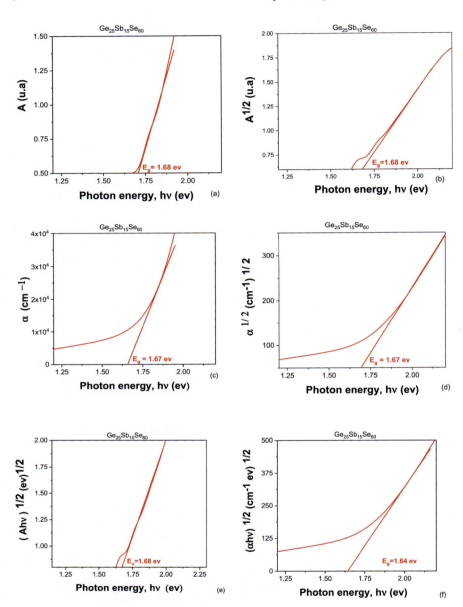

Fig. 5.5 The plot of A (**a**), $A^{\frac{1}{2}}$ (**b**), α (**c**), $\alpha^{\frac{1}{2}}$ (**d**) $Ah\nu^{\frac{1}{2}}$ (**e**), and $\alpha h\nu^{\frac{1}{2}}$ (**f**) versus $(h\nu)$ for $Ge_{25}Sb_{25}Se_{80}$

where A – is the absorptance, α – the absorption coefficient, $\widetilde{A}$, , $\widetilde{\alpha}$ are constants and E_u is the Urbach energy. The Urbach energy was deducted by the curves of $\ln A$ (Fig. 5.6) and $\ln \alpha$ (Fig. 5.7) as a function of photon energy $h\nu$. The reciprocal

5.1 Study of Optical and Chemical Properties

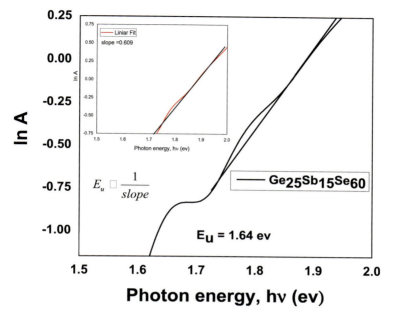

Fig. 5.6 The variation of Urbach energy deduced by ln A

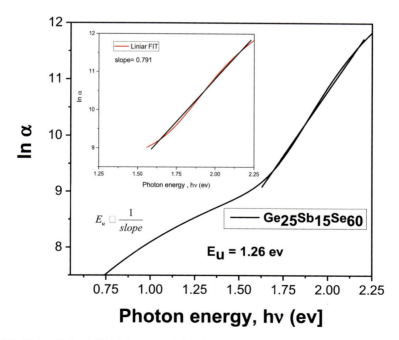

Fig. 5.7 The variation of Urbach energy deduced by ln α

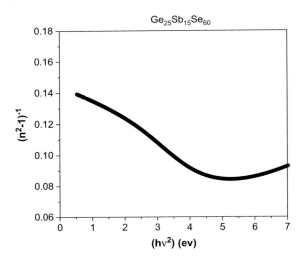

Fig. 5.8 The model's construction $(n^2 - 1)^{-1}$ against $(h\nu)^2$ for $Ge_{25}Sb_{15}Se_{60}$

slope of the linear fit, below the optical band-gap region, gives the value of the Urbach energy E_u.

Wemple-DiDomenico [12] (WD) model is used to calculate various dispersion parameters such as band-gap energy (E_g), oscillator energy (E_0), and dispersion energy (E_d). Accordingly, a Fig. 5.8 is constructed with $(n^2 - 1)^{-1}$ against $(h\nu)^2$. According to the single oscillator model, the data other dispersion of the refractive index can be evaluated using the equation:

$$n^2 = 1 + \frac{E_0 E_d}{E_0^2} - h\nu^2, \qquad (5.6)$$

where n is the refractive index, E_0 – oscillator energy, and E_d – dispersion energy.

Oscillator energy E_0 is calculated from the slope $1/E_0 E_d$ and dispersion energy E_d is calculated from the intercept E_0/E_d on y axis Fig. 5.9. The various dispersion parameters are also given in Table 5.1. E_g value is an average between the model based on the absorptance and absorption coefficient as a function of the photon energy $h\nu$.

Determination of High-Frequency Dielectric Constant

There are two procedures for obtaining and applying the high-frequency dielectric constant to the refractive index data, which are described as follows:

1. The first procedure describes the contribution of the free carriers and the lattice vibration modes of the dispersion.
2. The second procedure is based upon the dispersion arising from the bound carriers in an empty lattice.

Both models have used the refractive index estimation n, for the high-frequency dielectric constant ε_∞.

5.1 Study of Optical and Chemical Properties

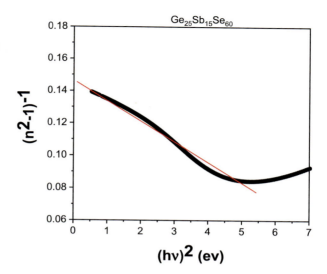

Fig. 5.9 Refractive index dispersion plots $(n^2 - 1)^{-1}$ versus $h\nu^2$ for Ge$_{25}$Sb$_{25}$Se$_{80}$

Table 5.1 Energetic parameters E_g, E_0, and E_d, for Ge$_{25}$Sb$_{15}$Se$_{60}$

Film composition	E_g (eV)	E_0 (eV)	E_d (eV)
Ge$_{25}$Sb$_{15}$Se$_{60}$	1.68	3.41	23.66

Zemel et al. [13] and Moss [14] developed a model to obtain the high-frequency dielectric using the relation:

$$\varepsilon_1 = \varepsilon_\infty - B\lambda^2, \tag{5.7}$$

where:

$$B = \frac{e^2 N}{4\pi^2 c^2 \varepsilon_0 m^*}, \tag{5.8}$$

ε_1 is the real part of dielectric constant, ε_∞ is the high frequency dielectric constant according to the first procedure, λ the wavelength, N the free charge carrier concentration, ε_0 is the permittivity of free space (8.854 × 10^{-12} F/m), m^* is the effective mass of the charge carrier and c the velocity of light. The real part of dielectric constants $\varepsilon_1 = n^2$ was calculated at different values of λ (Fig. 5.10). Extrapolating the linear part of this dependence to zero wavelength gives the value of ε_∞. The values of ε_∞ are given in Table 5.2 for sample Ge$_{25}$Sb$_{15}$Se$_{60}$.

The second procedure calculates the dielectric constant of material using the dispersion relation of the incident photon. The model of Moss [14] is applied in data corresponding to the wavelength range lying below the absorption edge of the material. The high-frequency dielectric constant can be calculated by using the following simple classical dispersion relation:

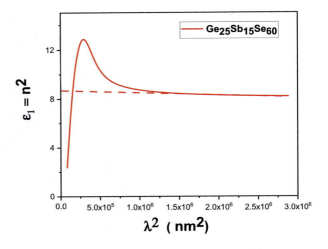

Fig. 5.10 Plots of ε_1 as a function of λ^2 for Ge$_{25}$Sb$_{15}$Se$_{60}$ sample

Table 5.2 The high-frequency dielectric constant for Ge$_{25}$Sb$_{15}$Se$_{60}$

Chalcogenide layers	From Fig. 5.10 ε_∞	From Fig. 5.11 ε_∞
Ge$_{25}$Sb$_{15}$Se$_{60}$	8.71	8.73

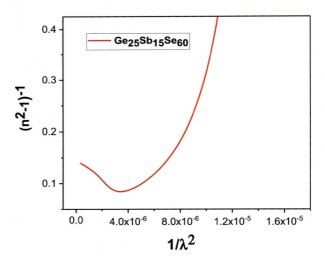

Fig. 5.11 $(n^2 - 1)^{-1}$ versus $(1/\lambda^2)$ curves for Ge$_{25}$Sb$_{15}$Se$_{60}$ sample

$$\frac{n_0^2 - 1}{n^2 - 1} = 1 - \left(\frac{\lambda_0}{\lambda}\right)^2 \qquad (5.9)$$

where n_0 is the refractive index at infinite wavelength λ_0, n-the refractive index and λ the wavelength of the incident photon. Plotting $(n^2 - 1)^{-1}$ against $1/\lambda^2$ (Fig. 5.12) showed a linear part. The intersection $(n^2 - 1)^{-1}$ axis is $(n_0^2 - 1)^{-1}$ and n_0^2 at λ_0 give ε_∞.

5.1 Study of Optical and Chemical Properties

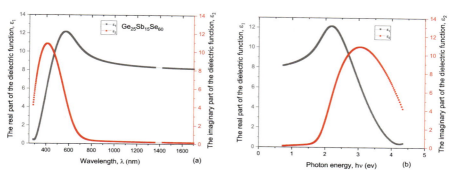

Fig. 5.12 ε_1 and ε_2 plots as a function of λ for Ge$_{25}$Sb$_{15}$Se$_{60}$ samples (**a**) representation of the dielectric function versus wavelength; (**b**) representation of the dielectric function versus photon energy

The dielectric constants ε_1, ε_2 are determined from the analytical dielectric complex function $\varepsilon = \varepsilon_1 + + i\varepsilon_2$, where ε_1 and ε_2 are calculated from the refractive index (n) and extinction coefficient (k).

$$\varepsilon_1 = n^2 - k^2 = \varepsilon_\infty - \left(\frac{e^2 N}{4\pi^2 c^2 \varepsilon_0 m^*}\right)\lambda^2, \tag{5.10}$$

$$\varepsilon_2 = 2nk = \left(\frac{\varepsilon_\infty \omega_p^2}{8\pi^2 c^2 \tau}\right)\lambda^3, \tag{5.11}$$

where:

- ε_1 is the real part,
- ε_2 is the imaginary part of the dielectric constant,
- ε_∞ is the high frequency dielectric constant,
- ω_p is the plasma frequency,
- τ is the optical relaxation time and can be calculated:

$$\tau = \left(\frac{\varepsilon_\infty - \varepsilon_1}{\omega_p \varepsilon_2}\right), \tag{5.12}$$

The values for ε_1 and ε_2 are present in Fig. 5.12.

The dissipation factor tan δ is described in Eq. (5.13) which is the power loss rate of the mechanical mode in a dissipative system [15].

$$\tan \delta = \frac{\varepsilon_2}{\varepsilon_1} \tag{5.13}$$

Figure 5.13 shows the dissipation factor plot tan δ against $h\nu$ for Ge$_{25}$Sb$_{15}$Se$_{60}$. As shown in Fig. 5.13, tan δ values display the same variation as the dielectric

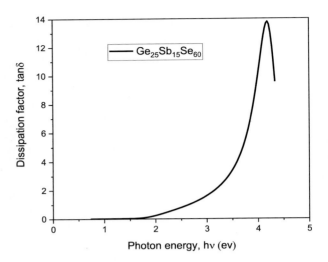

Fig. 5.13 Spectral dependence of the loss factor (tanδ) for $Ge_{25}Sb_{15}Se_{60}$

constants depicted in Fig. 5.12. The dissipation pick factor indicates interactions between the electrons; the interactions are also the origin of the peaks in the dielectric constant spectra.

Optical conductivity σ is used to detect the allowed inter-band optical transitions in the material. The complex optical conductivity $\widetilde{\sigma}$ is related to the complex dielectric constant $\widetilde{\varepsilon}$ [16] by the following relation:

$$\sigma_1 = \omega\varepsilon_2\varepsilon_0,$$
$$\sigma_2 = \omega\varepsilon_1\varepsilon_0 \qquad (5.14)$$

where ω is the angular frequency and ε_0 is the free space dielectric constant. The real and imaginary parts of the optical conductivity dependence of photon energy, $h\upsilon$, are shown in Fig. 5.14. It is seen in this figure that the real part increases with increasing frequency, while the imaginary part increases to energy exactly equal to the optical band-gap found by the method of Mott and Davis [17] in Fig. 5.4, then decreases.

From the values of ε_1 and ε_2 (Fig. 5.12), we obtained the complex electric modulus (Fig. 5.15) and the complex impedance (Fig. 5.16), which are given as follows:

$$M_e^* = \frac{1}{\widetilde{\varepsilon}} = M_{e_1} + iM_{e_2} = \frac{\varepsilon_1}{(\varepsilon_1{}^2 + \varepsilon_2{}^2)} + i\frac{\varepsilon_2}{(\varepsilon_1{}^2 + \varepsilon_2{}^2)},$$
$$Z^* = \frac{1}{i\omega C_0 \widetilde{\varepsilon}} = \frac{M_e^*}{i\omega C_0} = Z_1 + iZ_2, \qquad (5.15)$$
$$i^2 = -1,$$

5.1 Study of Optical and Chemical Properties

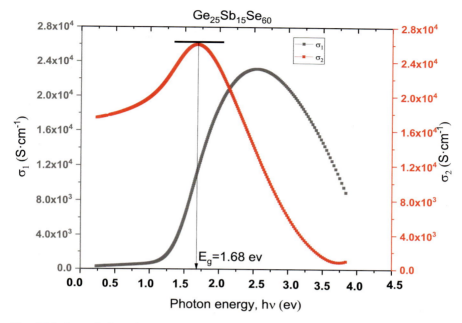

Fig. 5.14 Spectral dependence of the real and imaginary parts of optical conductivity for $Ge_{25}Sb_{15}Se_{60}$

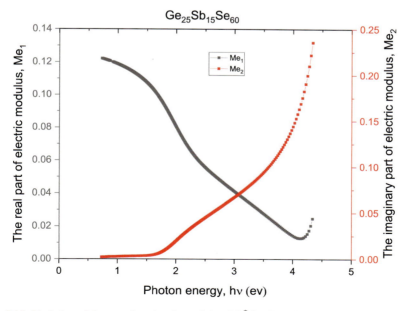

Fig. 5.15 Variation of the complex electric modulus, Me^* for $Ge_{25}Sb_{15}Se_{60}$

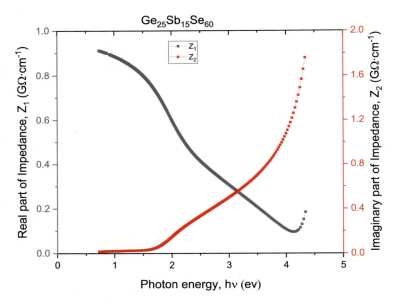

Fig. 5.16 Variation of the complex impedance, Z* for $Ge_{25}Sb_{15}Se_{60}$

where M_{e_1} and M_{e_2} are the real and imaginary parts of the electric modulus, respectively, Z_1 and Z_2 are the real and imaginary parts of the complex impedance, respectively, and $C_0 = \left(\frac{A}{d}\right)\varepsilon_0$ is the vacuum capacitance of the cell, A and d are the area and the thickness of chalcogenide layers, ε_0 is the free space dielectric constant.

The complex electric modulus is a decisive parameter to acquire information about the relaxation mechanism.

This method is widely used, and the results obtained give information about the optical and dielectric of the material properties. The information obtained can be correlated with the spectro-ellipsometry method.

5.2 UV-VIS-NIR and IR Spectrum Ellipsometry (SE)

It is a non-destructive indirect optical technique that allows the characterization of thin films, surfaces, and interfaces used to determine the thickness of thin films, optical constants, band-gap, electrical parameters (mobility, conductivity, carrier density) [18].

Principle of the Method

The physical principle of ellipsometry is illustrated in Fig. 5.17. The incident light interacts with the sample and then is reflected from it. The interaction of light on the sample produces a change in the polarization state of the light beam, from a linear to

5.2 UV-VIS-NIR and IR Spectrum Ellipsometry (SE)

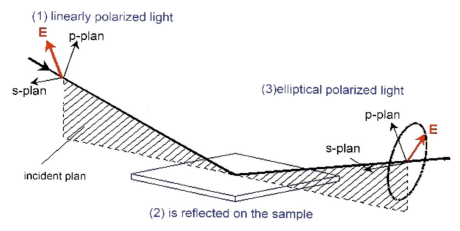

Fig. 5.17 The geometry of an ellipsometric measurement

an elliptical polarization. The change in polarization state is then measured by analyzing the lights reflected on the sample.

Ellipsometry does not directly measure the film thickness or optical constants. The values measured are Ψ, Δ, known as the SE parameters and are related to the ratio of Fresnel reflection coefficients, R_p and R_s, for p- and s- polarized light, from which the optical constants can be deduced:

$$\rho = \frac{R_p}{R_s} = \tan\psi e^{i\Delta} \tag{5.16}$$

This ratio is not sensitive to changes in the absolute intensity of the measurement beam. The analysis of ellipsometry data is generally performed by linear regression and the optical constants. The film structure is determined by minimizing the fit errors calculated with a specific mathematical function.

The linear regression analysis procedure [18] consists of the following stages:

(a) The first stage is represented by the acquisition of ellipsometric data by measuring the sample in a certain range of wavelengths and at certain angles of incidence. Figure 5.18 illustrates the data analysis procedure in SE.
(b) In the second stage, the optical model corresponding to the sample is constructed.

The analysis of the ellipsometry is performed by using an optical model. The model comprises an approximate structure of the samples, including the order of the layers, the optical constants of the material, and the thickness of each layer. When these are not known, it starts from a probable value. The fitting stage is represented by using a dedicated calculation program for generating data through various unknown parameters until finding an optimized set of parameters that would result in obtaining the best fit, as close as possible to the experimental data obtained from

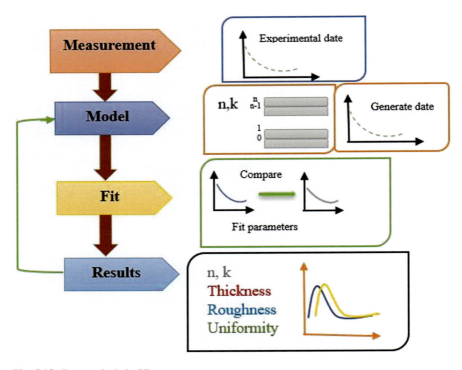

Fig. 5.18 Data analysis in SE

the measurement. The most used fitting parameters are thickness and optical constants. In some cases, the optical constants are listed in the literature [18–23].

The most used fitting algorithm is the Marquardt-Levenberg algorithm. The goal is to obtain the smallest differences between the measured and calculated values Ψ and Δ, i.e., the best fit. The quantification of the difference between these values is performed based on the mean square error (MSE), and thus the aim is to obtain the lowest possible values of this error. The formula for calculating the fitting error is as follows:

$$\mathrm{MSE} = \frac{1}{2N-M} \sum_{i=1}^{N} \left[\left(\frac{\Psi_i^{\mathrm{mod}} - \Psi_i^{\mathrm{exp}}}{\sigma_{\Psi,i}^{\mathrm{exp}}} \right)^2 + \left(\frac{\Delta_i^{\mathrm{mod}} - \Delta_i^{\mathrm{exp}}}{\sigma_{\Delta,i}^{\mathrm{exp}}} \right)^2 \right] = \frac{1}{2N-M} \chi^2 \quad (5.17)$$

where N is the number of pairs, M is the number of parameters that vary and present the standard deviation from the experimental data.

A low MSE error value shows a good fit of the theoretical model with the experimentally measured data. However, a low error value can also be obtained when too many parameters are set [24]. This leads to a result that is not unique.

5.2 UV-VIS-NIR and IR Spectrum Ellipsometry (SE)

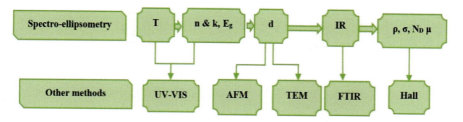

Fig. 5.19 Information obtained by spectroscopic ellipsometry (UV-VIS-NIR and Mid-IR), which can be obtained and verified by other methods

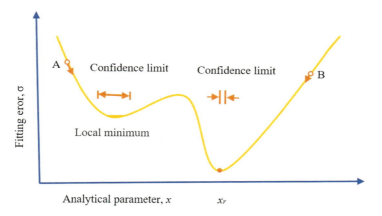

Fig. 5.20 Representation of linear regression analysis (x_r – the analytical solution obtained from the linear regression analysis)

To obtain a model that correctly describes the experimental data, as few fitting parameters as possible should be used. Various characterization techniques Fig. 5.19 are commonly used to verify the estimated structures of ellipsometry, including SEM – Scanning Electron Microscopy, TEM – Transmission Electron Microscopy, AFM – Atomic Force Microscopy [23, 24].

Also, the ellipsometric analysis performed at several angles provides a complete set of data necessary for the characterization of the sample. A diagram of the regression analysis is shown in Fig. 5.20. The analytical solution x_r is determined by minimizing the fit error σ as a function of the analytical parameter x.

Although σ_r is not a linear function, this analysis is called linear regression because the analytical parameter x is linear. Newton or Levenberg-Marquardt methods are used for linear regression analysis, and the analytical parameters used in the data analysis are determined simultaneously by applying these methods.

The confidence limit shows the accuracy of the regression analysis, and it becomes smaller when the absolute value of σ is small and its variation around x_r is steep. Unfortunately, the linear regression analysis depends a lot on the initial

106   5 Optical Analysis and Chemical Properties

values used in the analysis. Therefore, the initial values must be changed when the confidence limits or fitting errors are high [18–24].

In certain cases, there may be several reasons why a reasonable fit is not achievable or a low value of σ is obtained:

(a) The spectra (Ψ, Δ) are measured incorrectly;
(b) Dielectric functions used in data analysis are inadequate;
(c) The optical model is inadequate;
(d) The sample has a depolarizing effect.

The fit can be significantly improved by introducing a roughness to the model's surface or by inserting an interface layer. Also, when the optical properties of the thin film change in the direction of growth, the fit errors increase if the thin film is represented by a single layer (bulk) rather than represented by a multilayer model.

5.2.1 The Effective Medium Approximation (EMA) or EMT (Effective Medium Theory)

In some special cases [24], the optical functions of the thin film are fit by mediating a set of two or more optical functions, which correspond to the components of the material from which the thin layer is made. The problem is how to choose mediation. In the approximation of the effective medium, it is thus assumed that small particles of a material are suspended in the host matrix. In general, using the EMA model, the surface roughness layer is replaced by a single homogeneous and flat layer with an effective dielectric function (ε_{eff}). Bruggeman's EMA theory has been widely employed to calculate the effective dielectric function of the surface roughness layer and it takes the form:

$$\frac{\varepsilon_{\text{eff}} - \varepsilon_{\text{sample}}}{\varepsilon_{\text{eff}} + \gamma \varepsilon_{\text{sample}}} = \sum_i f_i \frac{\varepsilon_i - \varepsilon_{\text{sample}}}{\varepsilon_i + \gamma \varepsilon_{\text{sample}}} \tag{5.18}$$

where ε_{eff} is the dielectric function of the effective medium, $\varepsilon_{\text{sample}}$ is the dielectric function of the material sample, f_i is the fraction of constituents i, and γ is the factor relative to the shape of the inclusions. In the case of the previous equation, three models can be summarized:

1. Lorentz-Lorentz model: $\varepsilon_{\text{sample}} = 1$, the dispersion medium chosen is air. This is the first EMA theory and is based on the Clausius-Mossotti equation. It is thus assumed that we have an atomic scale mixture of individual constituents, so its application to real materials is limited because they tend to mix on a much larger scale.
2. Maxwell-Garnett model: $\varepsilon_{\text{eff}} = \varepsilon_1$, where the material is the one with the largest fraction of constituents. This material is useful when the inclusion fraction is

5.2 UV-VIS-NIR and IR Spectrum Ellipsometry (SE)

significantly smaller than the host material fraction, it is used, for example, in certain types of nanocrystals incorporated in the host matrix.

3. Bruggeman model (B-EMA): $\varepsilon_{eff} = \varepsilon_{sample}$ where the material is only the EMA dielectric function.

The surface roughness is fitted using the EMA Bruggeman model and is considered to be $f = 50\%$ voids and material. If the surface layer is very thick, it is necessary to incorporate several successive layers, each with different percentages of voids and material. If one of the materials is a complex dielectric function, it must be considered in the Bruggeman calculation. To avoid errors due to the non-use of these complex functions, the reparameterization of the Eq. (5.18) is used.

$$a = \sqrt{\frac{\varepsilon_1}{\varepsilon_2}}, \quad b = \frac{1}{4}\left[(3f_2 - 1)\left(\frac{1}{a} - a\right) + a\right]$$
$$c = b + \sqrt{b^2 + 0.5}$$

$$(5.19)$$

when the EMA of the dielectric function is given by the following equation:

$$\varepsilon_{eff} = c\sqrt{\sum_i \varepsilon_i} \tag{5.20}$$

In this case, by mathematical inversion, the measured Ψ, Δ can be converted directly to the optical constants n and k.

Conversely, utilizing the EMA model to calculate the SE parameters is a direct problem when the optical constants are known. Solving the inverse problem is more complex than solving the direct problem convenient for discussing the effects of the morphological parameters [25, 26].

5.2.2 Modeling Data

Nedelcu [6] studied the influence and the optical constant of the glassy composition based on chalcogenide $Ge_xSb_{40-x}Se_{60}$ alloys with $x = 12$, 25, and 30 at.%. The samples were measured in the UV-VIS-NIR spectral range 250–1700 nm, and at 50° angles of incidence, using ellipsometer variable angle spectroscopic ellipsometer (VASE) (J.A. Woollam Co.). The ellipsometric angles $\Psi(\lambda)$ Fig. 5.21a and $\Delta(\lambda)$ Fig. 5.21b were simulated by considering the samples as a two-layer optical system [substrate/first layer (film)/second layer (surface roughness)], Fig. 5.22, and using general oscillator mathematical models of corresponding software (data acquisition and analysis supplied by J.A. Woollam Co. Inc.). The fitting procedures were considered as a mixture of 50% material (film) and 50% voids (air) and were modeled by applying Bruggemann's effective medium approximation theory [27]. The optical constants were determined using Cauchy's equation in the spectral

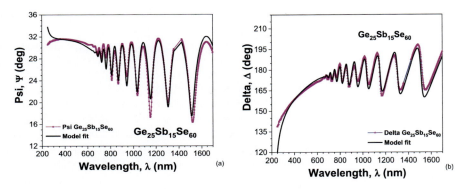

Fig. 5.21 Experimental data (blue line) and fitted data (red lines) (**a**) Ψ spectra and (**b**) Δ spectra for the Ge$_{25}$Sb$_{15}$Se$_{60}$

Fig. 5.22 The sample structure simulated by EMA

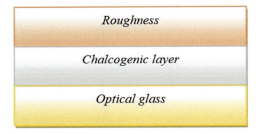

range of 400–1700 nm, where the films were expected to be transparent. Below 400 nm, a general oscillator approximation was applied to obtain the optical parameters. An iterative least-squares method was used for minimizing the difference (mean square error) between the experimental data and the theoretical one.

The optical constants Fig. 5.23 (refractive index *n* and extinction coefficient *k*) are determined from the SE data analysis in the spectral range 200–1700 nm. The results obtained by SE measurements are compared with the results calculated by UV-VIS-NIR spectroscopy.

The spectrophotometer's measurements were recorded on Lambda 950 Spectrophotometer with double beam and double monochromator at the room temperature, in the spectral range ultraviolet–visible-near-infrared (UV–VIS-NIR) with 266 nm/min a scanning speed for normal incidences. To correlate spectro-ellipsometry method with UV-VIS-NIR spectroscopy, the experimental data on the two methods were measured and compared in some conditions at normal incidence in the 250–1700 nm spectral range.

The total transmission $T(\lambda)$ [5] at normal incidence is given by:

5.2 UV-VIS-NIR and IR Spectrum Ellipsometry (SE)

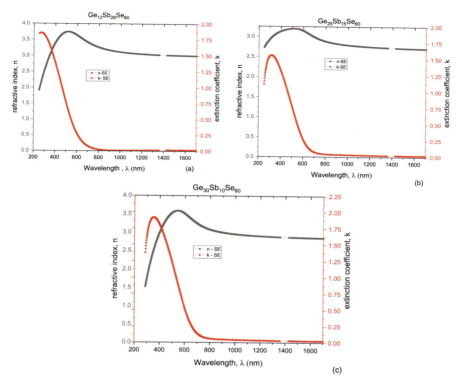

Fig. 5.23 Optical constant: refractive index n and extinction coefficient k, for (**a**) $Ge_{12}Sb_{28}Se_{60}$; (**b**) $Ge_{25}Sb_{15}Se60$; (**c**) $Ge_{30}Sb_{15}Se60$

$$T(\lambda) = \frac{16 n s_q^2 \tilde{x}}{(n+1)^3 \left(n+s_q^2\right) - 2(n^2-1)\left(n^2-s_q^2\right)\tilde{x}\cos\varphi + (n-1)^3\left(n-s_q^2\right)\tilde{x}^2},$$

$$\varphi = \frac{4\pi n d}{\lambda}, \qquad (5.21)$$

$$\tilde{x} = \exp(-\alpha d).$$

where n is the refractive index, s_q is the index of the quartz substrate, and the value is $s_q = 1.458$, d is thickness, φ is the phase difference between the direct and multiple reflected transmitted beams, $\tilde{x}$ is the absorbance and is the absorption coefficient. The thickness is determined from the deposition phase and experimental and is related in [3, 4, 6, 7].

Figure 5.24 shows the variation with respect to the transmittance wavelength for glassy systems $Ge_xSb_{40-x}Se_{60}$ alloys with compositions $x = 12, 25$, and 30 at.%, measured from both methods.

Manifacier et al. [28], following Swanepoel's method [5], propose calculating T_{max} and T_{min} by applying parabolic interpolation to the position of the experimental

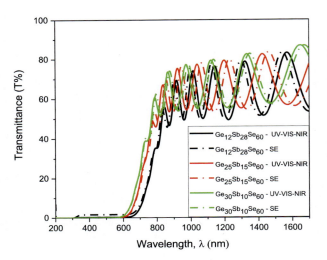

Fig. 5.24 Ge$_x$Sb$_{40-x}$Se$_{60}$ alloys with composition $x = 12, 25,$ and 30 at.%, measured from both methods (UV-VIS-NIR and SE)

peak on the envelope curves and setting the interference condition to $\cos\varphi = +1$, resulting in $T_{max}(\lambda)$. The expressions of T_{max} and T_{min} are the following:

$$T_{max}(\lambda) = \frac{16ns_q^2\widetilde{x}}{(n+1)^3\left(n+s_q^2\right) - 2(n^2-1)\left(n^2-s_q^2\right)\widetilde{x} + (n-1)^3\left(n-s_q^2\right)\widetilde{x}^2},$$

$$T_{min}(\lambda) = \frac{16ns_q^2\widetilde{x}}{(n+1)^3\left(n+s_q^2\right) + 2(n^2-1)\left(n^2-s_q^2\right)\widetilde{x} + (n-1)^3\left(n-s_q^2\right)\widetilde{x}^2}, \quad (5.22)$$

Considering the relationship above, the refractive index n can be calculated:

$$n = \sqrt{\frac{2s_q(T_{max}-T_{min})}{T_{max}T_{min}} - \frac{s_q^2-1}{2} + \sqrt{\frac{2s_q(T_{max}-T_{min})}{T_{max}T_{min}} - \frac{s_q^2-1}{2}^2} + s_q^2} \quad (5.23)$$

Figure 5.25 presents the refractive index n for Ge$_x$Sb$_{40-x}$Se$_{60}$ alloys with composition $x = 12, 25,$ and 30 at.% obtained by both methods of spectroscopy.

The thickness [3, 4, 6, 7] can be determined from two adjacent maxima (or minima) at λ' and λ'' using the relation:

$$d = \frac{\lambda'\lambda''}{2[\lambda' n''(\lambda'') - \lambda'' n'(\lambda')]} \quad (5.24)$$

The absorption coefficients are determined in the transmittance region of higher absorption using the relation:

5.3 Study of Chemical Properties

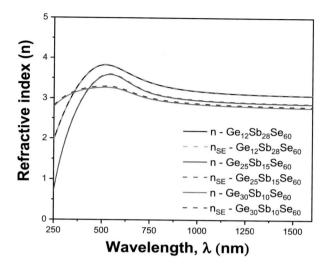

Fig. 5.25 The refractive index n obtained by both methods of spectroscopy

$$T(\lambda) = \frac{16ns^2\widetilde{x}}{(n+1)^3(n+s^2)} \quad (5.25)$$

$$\alpha(\lambda) = \frac{1}{d}\ln\left(\frac{16ns^2}{(n+1)^3(n+s^2)T(\lambda)}\right) \quad (5.26)$$

The extinction coefficient k is estimated from the values of the absorption coefficient $\alpha(\lambda)$ using the formula:

$$k = \frac{\alpha(\lambda)\lambda}{4\pi} \quad (5.27)$$

Figure 5.26 presents the extinction coefficient k for $Ge_xSb_{40-x}Se_{60}$ alloys with composition $x = 12$, 25, and 30 at.% obtained by both methods of spectroscopy.
The two methods provide complementary structural information.

5.3 Study of Chemical Properties

5.3.1 Infrared Spectroscopy-IR (FT-IR)

The molecular structure is based on many determinations of light absorption or scattering with various wavelengths by the analyzed sample, differentiated primarily by the energy (wavelength) of the radiation used. These methods can provide a wealth of information on the identity of molecules' structure and energy levels.

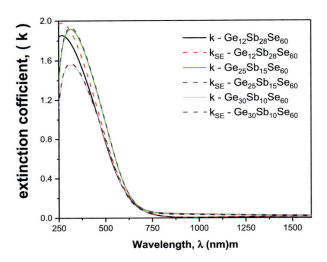

Fig. 5.26 The extinction coefficient k obtained by both methods of spectroscopy

An IR spectrum provides two types of information:

- energy or frequency of vibrational or roto-vibrational quantum transitions;
- the extent to which they absorb or emit radiation (intensity of this effect).

The transmittance of a sample of finite thickness τ is defined as the ratio between the radiant power transmitted by the sample ς_τ and the incident power ς_0

$$\tau(\sigma) = \frac{\varsigma_\tau(\sigma)}{\varsigma_0(\sigma)} \qquad (5.28)$$

where σ is content in the IR spectra range, the curve $\tau(\sigma)$ represents the vibrational spectrum of the substance and has values between 0 and 1 (or between 0% and 100% if expressed as a percentage).

There are two types of spectrometers: dispersive spectrometers and Fourier transform spectrometers. Fourier transform spectroscopy is based on interference, because interference between two radiation beams is only possible if the beams are coherent [29]. The easiest way to obtain a spectrum in this way is by using a Michelson interferometer and it is based on beam amplitude division of the wavefront from a lamellar network interferometer. The infrared beam from the source is transmitted to the interferometer. At the interferometer's output, the beam is sent to the sample, then to the detector. Compared to dispersive spectrometers, Fourier transform spectrometers have two main advantages:

- the acquisition time of the spectrum is considerably reduced;
- the energy of the beam is higher than that of a dispersive device, which gives the possibility to obtain spectra of very poorly absorbent samples [30].

5.3 Study of Chemical Properties

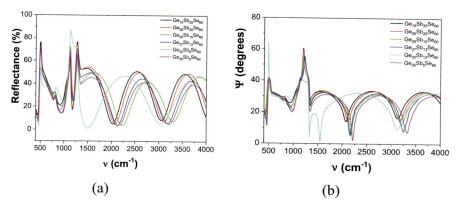

Fig. 5.27 The comparison of the two methods: (**a**) Spectral FTIR for $Ge_xSb_{40-x}Se_{60}$ (**b**) Spectral IRSE for $Ge_xSb_{40-x}Se_{60}$ [24]

Dulgheru [31] studied $Ge_xSb_{40-x}Se_{60}$ chalcogenide layers in the spectral range 400–4000 cm^{-1}, with a resolution of 4 cm^{-1} and 64 scans using Frontier Optica. The Kubelka-Munk is applied to convert the reflection spectrum into absorption units.

As mentioned above, the two methods (SE and UV-VIS-NIR) are complementary. We will show that they have the same behavior in the IR spectrum. Figure 5.27 compares the two methods: FTIR and IRSE $Ge_xSb_{40-x}Se_{60}$ [24].

The values of the vibrational modes are also found in the IRSE analysis, so the two types of measurements highlighted the same presence of the chemical bonds [24, 31].

5.3.2 Raman Spectroscopy

Raman spectroscopy is a method based on the inelastic scattering of an incident monochromatic light beam (a laser) on the optical phonons of the lattice crystal. Inelastic scattering occurs when the photon frequency changes as a result of interacting with the sample, specifically with the crystal lattice phonons. This involves absorption of the laser photons by the sample, followed by their re-emission. The retransmitted frequency is shifted; this represents the Raman effect. Raman shift does not provide information about vibrations, rotations, and other transition laws in molecules in a sample. The Raman effect is based on the molecular deformation in the electric field of the electromagnetic laser wave. Polarization deforms molecules resulting in vibrations, transitions, and oscillations of the phonon modes characteristic of certain structures and certain types of material [32, 33]. The light inelastic scattering phenomena on the molecule (Raman effect) was discovered in 1928. During the interactions between a photon and a molecule, exchange energy takes place: part of the photon's energy can pass over the molecule in the form of one or more quanta. Following this process, the molecule is excited on a higher vibrational level. The reverse process is also possible, in which part of the vibrational energy of the molecule is transferred to the photon, the molecule passing into a lower

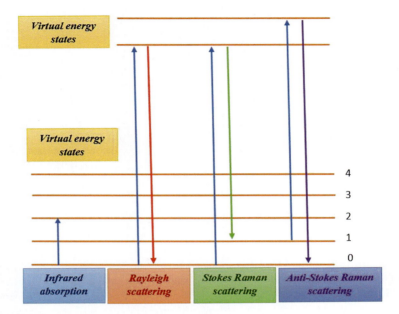

Fig. 5.28 Schematic energy level diagram showing the states involved in Raman signal

vibrational state. As a result of this inelastic collision, the energy of the emerging photon is different from that of the incident photon. The difference mirroring the energy position of the vibrational levels in the molecule. The Raman effect can also occur between rotational or electronic quantum levels. When light radiation of frequency ν_0 is transmitted over a sample, it can occur along with reflection and transmission of diffusion phenomena, either at the same frequency ν_0 (diffusion, Rayleigh) or at the frequency $\nu_R = \nu_0 + \nu_1$ (Raman diffusion).

The frequencies $\nu_R = \nu_0 - \nu_1$ are called Stockes, and the frequencies $\nu_R = \nu_0 + \nu_1$ are called anti-Stockes. Figure 5.28 presents the diagram of energy levels. The frequency changes (ν_1) are determined by the vibration frequencies of the chemical groups that make up the sample and are independent of the type of radiation used for stimulation. The Raman diffusion phenomenon is related to variations in electronic polarizability. Raman spectroscopy has several advantages over infrared spectroscopy: an important advantage is the possibility of studying aqueous solutions where water is a solvent, which causes great problems in the infrared.

Classical Theory

A static or low-frequency electric field induces a dipole moment in a molecule at high optical frequencies ($\sim 10^{15}$ Hz) because of the charged electrons and nuclei movement. The nuclei cannot respond rapidly, but the polarization of the electron distribution can occur enough to follow the field. For an isolated molecule, an oscillating radiation field of intensity **E** will induce a dipole moment **μ**

5.3 Study of Chemical Properties 115

$$\boldsymbol{\mu} = \alpha \mathbf{E} \tag{5.29}$$

where α is the molecular polarizability. The field $\mathbf{E}$ oscillates at the frequency ν_0 of the light, and the induced dipole will oscillate at this frequency:

$$E = E_0 \cos(2\pi \nu_0 t) \tag{5.30}$$

According to classical electromagnetic theory, any oscillating dipole will radiate energy. The light of frequency ν_0 is emitted in all directions (except that parallel to the dipole). Classical theory gives the total average intensity:

$$I = \frac{16\pi^4}{3c^3} \nu^4 \mu_0{}^2 \tag{5.31}$$

μ_0 – is the amplitude of μ, c is the velocity of light, the scattered radiation has the same frequency as $\mathbf{E}$. Inelastic or Raman scattering of light can be understood from modulation of the electron distribution, and the molecular polarizability, of the nuclei vibrations. For a diatomic molecule, α can be represented adequately by the first two terms of a power series in the vibrational coordinate ξ:

$$\alpha = \alpha_0 + \left(\frac{d\alpha_0}{d\xi_0}\right)\xi \tag{5.32}$$

The ξ oscillates at the vibrational frequency ν_v in the harmonic oscillator model according to the relation:

$$\xi = \widetilde{A} \cos(2\pi \nu_v t) \tag{5.33}$$

where $\widetilde{A}$ is the maximum amplitude of vibration. Equation 5.29 can be rewritten as:

$$
\begin{aligned}
\mu &= \alpha_0 E_0 \cos(2\pi \nu_0 t) + \frac{d\alpha_0}{d\xi_0}\widetilde{A}E_0 \cos(2\pi \nu_0 t)\cos(2\pi \nu_v t) \\
&= \alpha_0 E_0 \cos(2\pi \nu_0 t) + \frac{1}{2}\frac{d\alpha_0}{d\xi_0}\widetilde{A}E_0[\cos 2\pi(\nu_0 + \nu_v)t + \cos 2\pi(\nu_0 - \nu_v)t]
\end{aligned}
\tag{5.34}
$$

equation describes the amplitude modulation of the incident radiation by a simple harmonic oscillator, the first term is responsible for Rayleigh scattering at ν_0 while the second and third terms produce inelastic Raman scattering shifted by the frequency of the vibration, ν_v, to frequencies which are higher (anti-Stokes) and lower (Stokes), respectively, than the incident-light frequency. The frequency ν_0 usually corresponds to light in the visible region (typically the radiation from a laser). The Raman-shifted light then occurs in the visible region but with an intensity of 10^{-8} to 10^{-12} times that of the incident light. In the absence of such a resonance, the ratio of the Raman intensity of the anti-Stokes, I_A, and Stokes lines, I_S is:

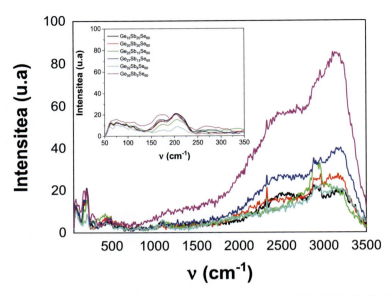

Fig. 5.29 FT-Raman spectra for Ge$_x$Sb$_{40-x}$Se$_{60}$ with composition $x = 15, 20, 25, 27, 32, 35$ at.%

$$\frac{I_A}{I_S} = \left(\frac{v_0 + v_v}{v_0 - v_v}\right)^4 e^{\frac{-hv_v}{KT}} \quad (5.35)$$

where k is Boltzmann's constant and T is the absolute temperature of the molecule. The Boltzmann exponential factor is dominant, and the anti-Stokes features are always much weaker than the corresponding Stokes lines [34].

Complementarity of Infrared and Raman Spectroscopes
The Infrared and Raman spectra are complementary, and the combination of the two techniques allows more information to be obtained. Selection rules prohibit certain vibrations, either infrared or Raman, but considerable differences in intensity can be observed for the same type of vibration. Infrared absorption, sensitive to variations in dipole moment, is intense especially for polar vibrators (C = O, C = X), while Raman diffusion, sensitive to variations in electronic polarizability, is especially indicated for the identification of highly polarizable groups (C = C, phenyl, etc.) [32–34].

Dulgheru [31] studied the Raman effect in chalcogenide glassy system GeSbSe. The structural analysis of the samples was performed with a Bruker Vertex 70 spectrometer equipped with RAM II mode and a Ge detector with variable power (1–500 mW). FT-Raman spectra were recorded between 50–3500 cm^{-1} with a scan speed of 512, a resolution of 4 cm^{-1}, and 500 mW power. Due to the fact that by Raman spectroscopy measurements can be made in the spectral range 50–350 cm^{-1} (Fig. 5.29), it was possible to determine the vibration modes of Sb that are in this region.

5.3 Study of Chemical Properties

Fig. 5.30 FT-IR spectra for Ge$_x$Sb$_{40-x}$Se$_{60}$

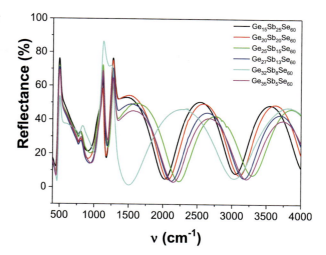

Non-stoichiometric compositions of Ge$_x$Sb$_{40-x}$Se$_{60}$ with composition $x = 15$, 20, 25, 27, 32, 35 at.% show characteristic bands of Se–Se bonds at 85 cm^{-1}. The bands at 64, 75, 95, and 110 cm^{-1} related the vibration of the Sb atoms in the Sb–Se pyramidal units. Their intensity decreases with decreasing Sb concentration in films [24]. The band from 121 cm^{-1} can be associated with the vibration of the Sb–Se connection, and the one from 170 cm^{-1} with the Ge–Ge connection from the Se$_3$Ge–GeSe$_3$ units. Other low-intensity bands were observed in the region of 250–350 cm^{-1} due to the Germanium content increase and can be attributed to vibrations Se–Se bonds and Ge–Ge, Ge–Sb bonds in the structural unit GeSe$_4$.

The absence of the 150 cm^{-1} band indicates the absence of Sb–Sb bonds in Se$_2$Sb–SbSe$_2$ structural units, [24] which is consistent with the low Sb concentration in the investigated films.

The vibrations (−Me−Me, Me−H−) in GeSe$_4$ tetrahedral units and certain impurities like SO−, SH, H$_2$O are responsible for the bands observed in the 400–4000 cm^{-1} region. These bands are also present in the results obtained from FTIR (Fig. 5.30) and IRSE analyses. The values assigned based on the literature data are present in Table 5.3.

Table 5.3 Raman vibrational modes as a Ge concentration [24]

Ge_{15} ν (cm^{-1})	Ge_{20} ν (cm^{-1})	Ge_{25} ν (cm^{-1})	Ge_{27} ν (cm^{-1})	Ge_{32} ν (cm^{-1})	Ge_{35} ν (cm^{-1})	Assigned bond
63.49	64.02	63.50	64.53	61.91	66.13	Sb (Se$_{1/2}$), GeS$_4$
76.56	75.51	76.03	76.04	75.02	75.51	Sb (Se$_{1/2}$), GeS$_4$
81.80	81.79	80.27	–	–	83.40	Ring Se$_8$
88.63	88.64	88.65	89.66	87.54	89.13	Ring Se$_8$
97.52	97.05	95.93	98.05	–	99.03	Ring Se$_8$
106.93	110.04	110.05	113.77	107.95	113.20	SbSe$_3$
120.54	121.57	122.10	–	121.10	–	GeSe$_2$, Sb–Se
127.90	129.45	128.41	–	127.35	131.06	Sb–Se
144.13	145.21	140.96	–	144.13	–	Sb–Se
169.76	169.27	–	170.86	167.15	170.85	Ge–Ge, Sb–Sb
–	–	176.62	–	178.16	178.73	Me–Se
204.39	207.04	205.41	204.37	206.41	202.77	Sb–Se
–	–	–	219.55	220.07	–	Ge–Se
–	238.92	248.31	–	242.06	236.30	Ge–Se, Se–Se
257.22	257.24	262.98	253.58	254.66	269.72	Ge–Se, Se–Se
–	–	272.46	273.48	275.60	278.72	Ge–Se, Se–Se
284.99	282.88	288.70	281.86	–	–	Ge–Se–Ge
–	–	–	296.52	293.37	294.48	GeSe$_4$
300.32	303.89	305.99	–	309.11	313.27	GeSe$_4$
318.01	323.76	322.15	316.96	327.92	–	GeSe$_4$
–	334.26	334.76	331.65	–	–	GeSe$_4$
342.60	343.95	–	346.21	349.93	347.88	GeSe$_4$

References

1. Mouler, J.F., *Handbook of X-ray photoelectron spectroscopy: a reference book of standard spectra for identification and interpretation of XPS data*, pg. 5–25, Editura Perkin- Elmer, USA, (1992).
2. Andrei Florin Danet ebook, *Analiza instrumentala- part I*, pg. 31–34 şi 99–104 Editura Universitatii din Bucuresti (2010).
3. N. Nedelcu, V. Chiroiu, C. Rugină, L. Munteanu, R. Ioan, I. Girip, C. Dragne, *Dielectric properties of GeSbSe glasses prepared by the conventional melt-quenching method*, Results in Physics, Volume 16, March 2020, 102856.
4. N. Nedelcu, V. Chiroiu, L. Munteanu, I. Girip, On the optical nonlinearity in the GeSbSe chalcogenide glasses, Materials Research Express, vol.7, No.6, 2020.
5. R.J. Swanepoel, *Determination of the thickness and optical constants of amorphous silicon*, J. Phys. E: Sci. Instrum. 16:1212–1222, 1980.
6. N. Nedelcu, V. Chiroiu, L. Munteanu, I. Girip, *Characterization of GeSbSe thin films synthesized by the conventional melt-quenching method*, Spectroscopy – IR Spectroscopy for today's Spectroscopists pg 22–33, vol 35, S3, August 2020.

References

7. N. Nedelcu, V. Chiroiu and L. Munteanu, Optimum GeSbSe layer design with respect to transmission and thickness, Optical Engineering 035109, 12 pgs. March 2021, Vol. 60(3).
8. N. Nedelcu, V. Chiroiu, L. Munteanu, I. Girip, C. Rugina, A. Lőrinczi, E. Matei, A. Sobetkii, *Design of highly transparent conductive optical coatings optimized for oblique angle light incidence*, Applied Physics A volume 127, Article number: 575 (2021).
9. J.I. Pankove, *Optical processes in Semiconductors*, New Jersey: Prentice-Hall p. 93, 1971.
10. J. Tauc 1974, *Amorphous and Liquid Semiconductors*, New York: Plenum Ch. 4, 1974.
11. Kumar NS, Bangera KV, Ananda C, Schivakumar GK 2013 Journal of Alloys and Compound 578 613–619.
12. Wemple S.H., DiDomenico, M. Jr., *Behavior of the Electronic Dielectric Constant in Covalent and Ionic Materials*, Phys. Rev. B 3, 1338 (1971).
13. Zemel, N., Jensen, J.D., Schoolar, R.B., *Electrical and Optical Properties of Epitaxial Films of PbS, PbSe, PbTe, and SnTe*, Phys. Rev. A, 140, 330 (1965).
14. Moss, T. S., *Optical Properties of Semiconductors*, Butterworth's Scientific Publication LTD., London, 245–265, 1959.
15. M. Fadel, S.A. Fayek, M.O. Abou-Helal, M.M. Ibrahim, A.M. Shakra, *Structural and optical properties of SeGe and SeGeX (X = In, Sb and Bi) amorphous films*, J. Alloy. Compd. 485(1), 604–609 (2009).
16. F. Bourguiba, A. Dhahri, T. Thari, K. Taibi, J. Dhahri, E. K. Hlil, *Structural, optical spectroscopy, optical conductivity and dielectric properties of $BaTi_{0.5}(Fe_{0.33}W_{0.17})O_3$ perovskite ceramic*, Bull. Mater. Sci., Vol. 39, No. 7, December 2016, pp. 1765–1774.
17. N.F. Mott and E.A. Davis, *Electronic processes in noncrystalline materials* (Oxford: Clarendon Press), 1979.
18. H. G. Tompkins, E. A. Irene, *Handbook of ellipsometry*, pg 4-70, William Andrew Publishing, Norwich, NY (2005).
19. J.N. Hilfiker, R.A. Synowicki, H.G. Tompkins, *Society of Vacuum Coaters*, 505/856-7188, (2008).
20. D. E. Aspnes, *Thin Solid Films*, 455–456, 3–13, (2004).
21. H. G. Tompkins, *A user's guide to ellipsometry*, pg 1–34, Academic Press. INC. (1993).
22. Critical Reviews of Optical Science and Technology, pg. 107–182, volume CR 72, Ed. Bellingham, Wash (1999).
23. K. Riedling, *Ellipsometry for industrial application*, pg. 51–72 Springer-Verlag-Wien (1988).
24. Dulgheru (Nedelcu) N, *Correlation of optical and morph-structural properties in chalcogenide compounds with applications in optoelectronics*, PhD thesis, Romanian Academy, 2019.
25. K. Megasari, E. Widianto, V. Efelina, K. Abraha, A. T. Seen Wee, A. Rusydi, I. Santoso, *Calculation of Dielectric Constant of Buffer Layer Graphene on SiC Measured by Spectroscopy Ellipsometry using Gauss-Newton Numerical Inversion Method*, AIP Conference Proceedings 1755, 150014 (2016)
26. B. Fodor, P. Kozma, S. Burger, M. Fried, P. Petrik, *Effective medium approximation of ellipsometric response from random surface roughness simulated by finite-element method*, Thin Solid Films 617 (2016) 20–24.
27. D.A.G. Bruggeman, *Berechnung verschiedener physikalischer Konstanten von heterogenen Substanzen. I. Dielektrizitätskonstanten und Leitfähigkeiten der Mischkörper aus isotropen Substanzen*, Ann. Phys. 24 (1935) 636–644.
28. Manifacier, J. C., Gasiot, J., and Fillard, J. P., *A simple method for the determination of the optical constants n,k and the thickness of a weakly absorbing thin film*, Phy. E: Sci. Instrum. 9 (1976) 1002–1004, 1976.
29. L. I. David, C.E. Cristea, O. Cozar, L. Găină, *Identificarea structurii moleculare prin metode spectroscopice*, pg.1–25, Editura Universității din Bucuresti (2004).
30. Brian C. Smith, *Fundamentals of Fourier Transform Infrared Spectroscopy*, pg 1–53, Ed. CRC Press (2011).

31. N. Dulgheru, M. Gartner, M. Anastasescu, M. Stoica, M. Nicolescu, H. Stroescu, I. Atkinson, V. Bratan, I. Stanculescu, A. Szekeres, P. Terziyska, M. Fabian, *Influence of compositional variation on the optical and morphological properties of GeSbSe films for optoelectronics application*, Infrared Physics and Technology 93 (2018) 260–270.
32. G. Dent, E. Smith, *Modern Raman Spectroscopy*, pg 3–27, Ed. John Willy&Sons LTD (2005).
33. J.R Ferraro, ebook, *Introductory Raman Spectroscopy*, pg. 1–61, Elsevier Science (2002).
34. S. Garoff, B. Liokkala, *Characterizing, molecular vibrations using Raman Spectroscopy*, Carnegie Mellon University, 2010.

Index

A

Ag, 53
Al, 54
Antireflection coating, 90
Atomic force microscopy (AFM), 66–69
Auger Electron Spectroscopy (AES), 72
Auger electrons, 78
Automated crystallography (ACT), 81–84
Auxiliary phases, 34

B

Backscattered Electron SEM Imaging, 72
Boltzmann exponential factor, 116
Boltzmann's equation, 42
Bragg angle, 80
Bragg's laws, 85
Bragg's conditions, 74
Bruggeman calculation, 107
Bruggeman model (B-EMA), 107
Bruggeman's EMA theory, 106

C

CaF_2, YF_3, 54
Capacitive method, 56
Cauchy equation, 6
Cauchy formula, 5
Cauchy's equation, 107
Characterization methods
    microscopy techniques, 66
        atomic force microscopy, 66–69
        operation description of, 78

        scanning electron microscopy, 69–74
        transmission electron microscopy,
            74–86
    neutron diffraction, 63–66
    structural and morphological analysis,
        62, 63
    thin layers, methods, 61
Charge-coupled device (CCD), 74
Chemical vacuum deposition method
    (CVD), 44, 45
Coating thermal spray materials, 37
Condenser lenses, 76
Conical dome, 20, 21
Crucibles, 51
Cryogenic pumps, 49
Cu, 54

D

Debye equation, 63
Detonation gun spraying, 39
Dispersion, 4

E

Effective medium approximation
    (EMA), 106, 107
Effective medium theory (EMT), 106, 107
Elastically scattered electrons, 78
Electric arc spraying process, 40, 41
Electrochemical methods, 43, 44
Electron backscatter diffraction (EBSD), 73–75
Electron beam, 52

© The Editor(s) (if applicable) and The Author(s), under exclusive license to
Springer Nature Switzerland AG 2023
N. Nedelcu, *Thin Films*, https://doi.org/10.1007/978-3-031-06616-0

121

Index

Electron energy loss spectroscopy, 79
Electrons, 75
Ellipsometric angles, 107
Ellipsometry, 89
Energy dispersive X-ray spectroscopy (EDS), 71–73
Epitaxial growth, 34
Evaporation
    materials for, 53, 54
Evaporation devices, 49–53

F
Focused ion beam (FIB) technique, 80

G
Gas phases, 34
Ge, 54
Geometric coefficient, 18
Ghetto-ionic/magneto-ionic pumps, 48
Gravimetric method, 58

H
Harmonic oscillator model, 115
High-frequency dielectric constant, determination of, 96–102
High-frequency plasma spraying, 34
High-resolution transmission electron microscopy (HRTEM), 81

I
Inductive method, 56, 57
Inelastically scattered electrons, 79
Infrared spectroscopy-IR (FT-IR)
    chemical properties, 111–113
Inhomogeneities
    types of, 7, 8
Intermediate lenses, 77
Ion beam, 52
IR spectrum ellipsometry (SE), 89, 102
    EMA, 106, 107
    modeling data, 107–111
    principle of method, 102–106

J
Joule effect, 49

K
Kikuchi bands, 74, 79
Kikuchi lines, 79, 80

L
Lambda spectrophotometer, 91
Laser ablation
    thermal evaporation by, 36, 37
Laser beam, 36
Laser beam heating method, 52
Laser pulse energy, 36
Least-squares method, 108
Lorentz-Lorentz model, 106

M
Maxwell-Garnett model, 106
Mean square error (MSE), 104
MgF_2, 54
Micro-characterization, 69
Microscopy techniques, 66
    atomic force microscopy, 66–69
    scanning electron microscopy, 69–72
        electron backscatter diffraction, 73, 74
        energy dispersive X-ray spectroscopy, 71–73
    transmission electron microscopy, 74–77
        applications, 80
        Auger electrons, 78
        automated crystallography, 81–85
        electronic interactions, in sample analyzed, 77, 78
        high-resolution transmission electron microscopy, 81
        Kikuchi lines, 79, 80
        secondary electrons, 78
        selected area electrons diffraction, 84–86
        small thickness samples, interactions in, 78, 79
Microstructure, 75
Miller indicators, 74
MSE error value, 104

N
Neutron diffraction (ND), 63–66
Newton/Levenberg-Marquardt methods, 105
Nondispersive medium, 5
Noran System Detector, 73

O
OpenFilters software, 90
Optical absorption coefficient, 91
Optical band-gap, 93

Index

Optical coating
  uniformity of, 9, 13, 15–17
    plane support, 17, 18
    pyramidal and conical dome, 20–22
Optical conductivity, 90, 100
Optical materials, 2
Oscillator energy, 96
Oxides, 53

P

Peak detection, 81
Pegasus system, 74
Phase-contrast images, 81
Planetary geometry, 20
Planetary system geometry, 23–26
  aspherical surfaces in, 9–17
Plasma jet, 41
Plasma spraying, 41, 42
PLD general assembly, 37
Polarization deforms molecules, 113
Projector lenses, 77
Pulsed laser, 52
Pyramidal dome, 20–22

R

Raman diffusion phenomenon, 114
Raman effect, 113
Raman scattering, 115
Raman spectroscopy
  chemical properties, 113, 114
    classical theory, 114–116
      Infrared and Raman spectroscopes,
        complementarity of, 116–118
Refractive index, 4
Resonant quartz method, 57, 58
Reverse Monte Carlo (RMC) method, 65

S

Scanning electron microscopy (SEM), 69–72
  electron backscatter diffraction, 73–75
  energy dispersive X-ray spectroscopy,
    71–73
Scattered electrons, 77, 78
Scattered radiation, 63
Scattering, 85
Secondary electrons, 78
Selected area electrons diffraction
  (SAED), 84–86
Sellmeier coefficient, 6
Sellmeier's equation, 6

Silicone oil pumps, 49
SiO, 53
SiO_2, 53
Small thickness samples, interactions in, 78, 79
Spectro-ellipsometry method, 102
Spectroscopy, 90
Spherical dome, 19, 20
Static/low-frequency electric field, 114
Swanepoel method, 92
Swanepoel model, 89

T

Ta_2O_5, 53
Thermal evaporation
  by laser ablation, 36, 37
  thin layer obtaining by, 34, 35
Thermal spraying
  thin layers by, 37–39
    detonation gun spraying, 39, 40
    electric arc spraying process, 40, 41
    plasma spraying, 41, 42
Thickness measurement methods, 54, 55
  capacitive method, 56
  gravimetric method, 58
  inductive method, 56, 57
  resonant quartz method, 57, 58
Thin conductor, 49
Thin layer
  methods for, 33, 34, 61
  by thermal evaporation method, 34, 35
  by thermal spraying, 37–39
    detonation gun spraying, 39, 40
    electric arc spraying process, 40, 41
    plasma spraying, 41, 42
TiO_2, 53
Total transmission, 108
Transmission electron microscopy
  (TEM), 74, 75
  applications, 80
  Auger electrons, 78
  automated crystallography, 81–84
  electronic interactions, in sample
    analyzed, 77, 78
  high-resolution transmission electron
    microscopy, 81
  Kikuchi lines, 79, 80
  operation description of, 76, 77
  secondary electrons, 78
  selected area electrons diffraction, 84–86
  small thickness samples, interactions
    in, 78, 79
Turbo-molecular pumps, 48

U

Ultraviolet–visible-near-infrared (UV–VIS-NIR), 108
Uniformity screens, 26–29, 31
Urbach empirical rule, 93
Urbach energy, 95
Urbach tail, 93
UV-VIS-NIR, 89, 102
  EMA, 106, 107
  modeling data, 107–111
  principle of method, 102–106
UV-VIS spectroscopy, 90, 91, 93–96
  high-frequency dielectric constant, determination of, 96–102

V

Vacuum pumps, 47, 48
Vacuum thermal evaporation method, 34
Vacuum thin film deposition installations, 47–49
Vacuum valves, 47

Variable angle spectroscopic ellipsometer (VASE), 107
Virtual source, 76

W

Wobbler, 76

X

X-ray diffraction, 85
X-ray emission, 72
X-ray spectrometer, 73

Y

Y_2O_3, 53

Z

ZnS, 54
ZnSe layer, 73
ZrO_2, 53

Printed in the United States
by Baker & Taylor Publisher Services